The Silviculture of Mahogany

The Silviculture of Mahogany

J.E. Mayhew
and
A.C. Newton

Institute of Ecology and Resource Management
University of Edinburgh
UK

CABI *Publishing*

CABI *Publishing* - a division of CAB INTERNATIONAL

CABI *Publishing*
CAB INTERNATIONAL
Wallingford
Oxon OX10 8DE
UK

CABI *Publishing*
10 E 40th Street
Suite 3203
New York, NY 10016
USA

Tel: +44 (0)1491 832111
Fax: +44 (0)1491 833508
Email: cabi@cabi.org

Tel: +1 (212) 481 7018
Fax: +1 212 686 7993
Email: cabi-nao@cabi.org

©CAB INTERNATIONAL 1998. All rights reserved. No part of this publication may be reproduced in any form or by any means, electronically, mechanically, by photocopying, recording or otherwise, without the prior permission of the copyright owners.

A catalogue record for this book is available from the British Library, London, UK.

Library of Congress Cataloging-in-Publication Data
Mayhew, J. E. (John E.)
　　The silviculture of mahogany (Swietenia macrophylla) / J.E Mayhew & A.C. Newton.
　　　　p.　cm.
　　Includes bibliograpical references and index.
　　ISBN 0-85199-307-9
　　1. Honduras mahogany. 2. Tree farms. 3. Forests and forestry.
　I. Newton, Adrian C. II. Title.
　SD397.H67M39　1998
　　634.9'7377--dc21　　　　　　　　　　　98-34213
　　　　　　　　　　　　　　　　　　　　　　　CIP
ISBN 0 85199 307 9

Printed and bound in Great Britain by Biddles Ltd, Guildford and King's Lynn

Contents

ACKNOWLEDGEMENTS ix

1 INTRODUCTION 1

2 DESCRIPTION OF THE SPECIES 3
2.1 TAXONOMY AND DISTRIBUTION OF *SWIETENIA* 3
2.2 GENERAL DESCRIPTION OF *S. MACROPHYLLA* 4
2.3 ECOLOGICAL RANGE 7
2.4 REPRODUCTIVE BIOLOGY 8
2.5 REGENERATION ECOLOGY 9
   2.5.1 Seed dispersal 9
   2.5.2 Conditions required for germination 10
   2.5.3 Conditions required for seedling establishment 10
   2.5.4 Natural regeneration and the role of disturbance 12
   2.5.5 Associated species 14
2.6 SUMMARY 15

3 MAHOGANY AS A PLANTATION SPECIES 17
3.1 ORIGIN OF SEED USED TO ESTABLISH MAHOGANY PLANTATIONS 17
3.2 ATTEMPTS TO IMPROVE QUALITY OF PLANTING STOCK 24
   3.2.1 Provenance trials and progeny tests 24
   3.2.2 Genetic improvement 25
3.3 SUMMARY 27

4 SEED PRODUCTION 27
4.1 SEED COLLECTION 27
4.2 SEED PREPARATION 29
4.3 SEED VIABILITY 30
   4.3.1 Viability of fresh seed 30
   4.3.2 Viability of stored seed 31
4.4 SUMMARY 33

5 NURSERY TECHNIQUES 34
5.1 SOWING 34
   5.1.1 Sowing in beds 34
   5.1.2 Sowing in containers 35

5.1.3 Sowing in trays 35
5.1.4 Optimal sowing depth and seed orientation 35
5.1.5 Germination 36
5.2 VEGETATIVE PROPAGATION 37
5.2.1 Propagation by leafy cuttings 37
5.2.2 In vitro micropropagation 38
5.3 MAINTENANCE 38
5.3.1 Shading 38
5.3.2 Root pruning 39
5.3.3 Controlling pests and diseases 40
5.4 SEEDLING GROWTH AND RECOMMENDED SIZE FOR PLANTING OUT 40
5.5 PREPARATION OF BARE-ROOTED STOCK FOR PLANTING OUT 43
5.5.1 Leaf stripping 44
5.5.2 Cutting back 44
5.5.3 Storage 45
5.6 SUMMARY 45

6 SITE SELECTION 47

6.1 CLIMATE AND ALTITUDE 47
6.2 SOIL TYPE 48
6.3 TOPOGRAPHY 53
6.4 INDICATIVE SPECIES 54
6.5 SUMMARY 54

7 PLANTATION ESTABLISHMENT 54

7.1 DIRECT SEEDING 54
7.2 PLANTING IN THE OPEN 57
7.2.1 Pure mahogany plantations 57
7.2.2 Mixed mahogany plantations 58
7.2.2.1 Agroforestry systems 58
7.2.2.2 Forestry systems 66
7.2.2.3 Silvipastoral systems 67
7.3 PLANTING IN NATURAL FOREST 69
7.3.1 Conversion line planting 69
7.3.1.1 Removing the overstorey 72
7.3.1.2 Cutting lines 73
7.3.1.3 Planting density 74
7.3.1.4 Problems encountered with conversion line planting 76
7.3.2 Conversion underplanting 76
7.3.3 Enrichment planting 78
7.4 SUMMARY 79

8 PLANTATION MAINTENANCE 80

8.1 WEEDING 80
8.2 CLEANING 83
8.3 FERTILISING 83
8.4 PRUNING 84
8.5 THINNING 86
8.5.1 Thinning pure mahogany plantations 86

8.5.2 Thinning mixed mahogany plantations 94
8.5.3 Thinning mahogany line plantings 94
8.6 ROTATION LENGTH 95
8.7 SUMMARY 96

9 GROWTH AND YIELD 97

9.1 AVAILABLE GROWTH DATA 97
9.1.1 Limitations and assumptions 98
9.1.2 Graphical relationships 100
9.2 PREDICTIONS OF GROWTH AND YIELD 103
9.2.1 Height predictions 104
9.2.1.1 Mean height 104
9.2.1.2 Dominant height and site index curves 105
9.2.1.3 Relationship between mean and dominant height 108
9.2.2 Diameter predictions 108
9.2.2.1 Mean diameter 108
9.2.2.2 Dominant diameter 112
9.2.3 Yield predictions 113
9.3 ESTIMATION OF SINGLE TREE VOLUME 114
9.4 SUMMARY 120

10 TIMBER QUALITY 122

10.1 EFFECTS OF SITE AND CLIMATE ON TIMBER QUALITY 124
10.2 SUMMARY 124

11 SHOOT BORER CONTROL 125

11.1 SILVICULTURAL CONTROL OF *HYPSIPYLA* SPP. 127
11.1.1 Proposed mechanisms for silvicultural control 127
11.1.1.1 Location 128
11.1.1.2 Susceptibility 128
11.1.1.3 Tolerance 129
11.1.1.4 Natural enemies 129
11.1.2 Techniques for silvicultural control pre-planting 130
11.1.2.1 Provenance selection 130
11.1.2.2 Site selection 130
11.1.2.3 Plantation design 130
11.1.3 Techniques for silvicultural control post-planting 135
11.1.3.1 Weeding and cleaning 135
11.1.3.2 Fertilising 135
11.1.3.3 Pruning 135
11.1.3.4 Thinning 136
11.1.4 Evaluation of reported experiences and recommendations 136
11.2 CHEMICAL CONTROL OF *HYPSIPYLA* SPP. 137
11.3 BIOLOGICAL CONTROL OF *HYPSIPYLA* SPP. 137
11.3.1 Classical biocontrol 138
11.3.2 Augmentative biocontrol 139
11.3.3 Microbial control 139
11.4 SUMMARY 140

12 PROTECTION 141

12.1 CONTROL OF PESTS AND DISEASES 141
    12.1.1 Ambrosia beetles 141
    12.1.2 Termites 143
    12.1.3 Other pests and diseases 145
12.2 PROTECTION FROM HURRICANES 149
    12.2.1 Comparative resistance to hurricane damage 149
    12.2.2 Limitation of hurricane damage 149
12.3 PROTECTION FROM FIRE 150
    12.3.1 Comparative resistance to fire damage 150
    12.3.2 Limitation of fire damage 151
12.4 SUMMARY 151

13 SILVICULTURAL SYSTEMS 152

13.1 MANAGEMENT OF MAHOGANY IN NATURAL FORESTS 153
    13.1.1 Commercial scale management 153
    13.1.2 Research scale management 155
13.2 SILVICULTURAL SYSTEMS IN PLANTATIONS 158
    13.2.1 Natural regeneration in mahogany plantations 158
    13.2.2 Systems utilising natural regeneration of mahogany 160
        13.2.2.1 A single tree selection system in Sri Lanka 160
        13.2.2.2 A modified uniform shelterwood system in Sri Lanka 162
        13.2.2.3 A group selection system in St. Lucia 164
        13.2.2.4 Application of silvicultural systems in other countries 166
13.3 SUMMARY 167

14 CONCLUSIONS 169

PERSONAL COMMUNICATIONS 172

REFERENCES 175

APPENDICES 199

APPENDIX I THINNING REGIMES FOR MAHOGANY STANDS 199
APPENDIX II STOCKING OF MAHOGANY STANDS BASED ON DBH-CROWN DIAMETER
    RELATIONSHIPS 201
APPENDIX III DATA FROM SINGLE INVENTORIES OF MAHOGANY STANDS 203
APPENDIX IV DATA FROM CONTINUOUS INVENTORIES OF MAHOGANY STANDS 209
APPENDIX V YIELD TABLES FOR MAHOGANY STANDS 213
APPENDIX VI VOLUME TABLES FOR MAHOGANY TREES 217

INDEX 220

Acknowledgements

The authors gratefully acknowledge the invaluable assistance of institutions and individuals in the following countries:

Belize
Forest Planning and Management Project (FPMP), Overseas Development Administration (ODA): Neil Bird.
Forest Department: Oswaldo Sabido, Francisco Villafranco, Richard Ennion.
Programme for Belize: Roger Wilson.

Costa Rica
Centro Agronómico de Investigación y Enseñanza (CATIE): Jonathan Cornelius, Carlos Navarro Pereira, Scott Stanley, Steve Gretzinger, Paul Martins, Philip Shannon, Eduardo Somarriba, library staff.

Fiji
Silvicultural Research Division, Department of Forestry: Kuldeep Singh, Tevita Evo, Tom Kubuabola, Madu Kamath.
Management Division, Department of Forestry: Rupeni Anise, Netani Moce.
Wood Technology Division, Department of Forestry: Manoj Chand.
Southern Division, Department of Forestry: Lavisai Seroma, Waisea Tuicicia.

Honduras
Lancetilla Botanic Gardens: Melvin Cruz.
Proyecto de Conservación y Mejoramiento de Recursos Forestales, ODA-CONSEFORH: Ernesto Ponce, Joss Wheatley.
Proyecto Desarollo del Bosque Latifoliado: Denis Buteau, Raphael Meza.
Cooperación para el Desarrollo de Paises Emergentes, Cooperativa COATLAHL: Filippo Del Gatto.

Malaysia
Forest Research Institute of Malaysia (FRIM): Simithiri Appanah, Marzalina Mansor.
Yayasan Sabah: Chan Hing Hon, John Tay, James Kotulai, Kazuma Matsumoto.

Mexico
Quintana Roo Forestry Management Project, ODA: Zafar Ul Hassan, Ana Malos.
Plan Piloto, Deutsche Gesellschaft für Technische Zusammenarbeit (GTZ): Daniel Gonzalez, Felipe Sanchez Román.

Instituto Nacional de Investigaciones Forestales y Agropecuarias (INIFAP): Conrado Parraguirre.
Las Caobas Ejido: Salvador Gutierres and Ramón Cabral.

Philippines
Ecosystems Research and Development Bureau (ERDB), Department of Environment and Natural Resources (DENR): Celso Diaz, Felizardo Virtucio, Norma Pablo, Eraneo Lapiz, Aida Lapiz, Evangeline Castillo.
Forest Management Bureau (FMB), DENR: Jose Malvas, Jr., Wilfredo Malvar.
Environmental and Resources Research Division (ERRD), DENR: Angelito Valencia, Henrico Patricio.

Puerto Rico
International Institute of Tropical Forestry (IITF), United States Department of Agriculture: John Francis, Peter Weaver, Frank Wadsworth, Julio Figueroa Colón, IITF library staff.

St. Lucia
Forestry Department, Ministry of Agriculture, Lands, Fisheries and Forests: Brian James, Lyndon John, Michael Andrew, Michael Bobb, Peter Vidal, Monica Bodley.

Solomon Islands
Forest Research Station, Ministry of Forests, Environment and Conservation: Simeon Iputu.
Kolombangara Forest Products Ltd. (KFPL): Wayne Wooff, Moray Isles, Paul Speed, Xavier Viuru, John Lilo.

Sri Lanka
Forest Department: S. Thayaparan, S. Wickramasinghe.
Kumbalpola Forest Research Station, Forest Department: N.D.R. Tilakeratne, Weerawardene.
Forest Management and Plantation Project (FORMP), ODA: Jim Sandom, John Pallot, David Danbury, Julian Gayfer.

United Kingdom
Institute of Ecology and Resource Management (IERM), Edinburgh University: Douglas Malcolm, Carrie Hauxwell, Jim Wright, Darwin Library staff.
Institute of Terrestrial Ecology (ITE), Edinburgh: Amanda Gillies, Allan Watt.
Oxford Forestry Institute (OFI), Oxford University: OFI library staff.
Forestry Research Programme (ODA/DFID): John Palmer, Howard Wright, Anne Bradley.

The authors would also like to thank all those who provided contacts (Tang Hon Tat, Peter Neil, John Palmer, Jim Coles, Mike Shaw, Jon Heuch, Tim Nolan); those who sent information (Thomas Wormald, Malaki Iakapo, Iro Wanefaia, Jerry Vanclay, John Fox, Hennig Flachsenberg, Nina Marshall, Ted Gullison, Laura Snook); those who reviewed draft documents (including Peter Savill, Howard Wright, Jim Sandom, Jim Ball); and to Dr Peter Baker, who provided the section on biological control of *Hypsipyla*.

This publication is an output from a research project funded by the Department for International Development of the United Kingdom. However, the Department for International Development can accept no responsibility for any information provided or views expressed. Project code R6351, Forestry Research Programme. The financial support of DFID is gratefully acknowledged.

Contact details

John Mayhew and Adrian Newton
Institute of Ecology and Resource Management,
University of Edinburgh,
Darwin Building,
Mayfield Road,
Edinburgh EH9 3JU,
United Kingdom.

e-mail:
john.mayhew@ed.ac.uk
A.Newton@ed.ac.uk

Chapter One

Introduction

It is now over 30 years since the classic work on mahogany (*Swietenia macrophylla*), entitled *Mahogany of Tropical America: Its Ecology and Management,* was produced by Lamb (1966). His book focuses on the ecology and management of mahogany in natural forests, with only two out of the nine chapters dedicated to plantations. Since Lamb's book, there have been two important phases of research on mahogany. The first took place in the 1970s at the Centro Agronómico de Investigación y Enseñanza (CATIE) in Costa Rica, where considerable efforts were made to tackle the problem of the mahogany shoot borer, the main factor limiting cultivation of mahogany in plantations. Although a considerable body of research information was produced, few practical solutions were proposed and the value of mahogany as a plantation species was widely questioned. More recently, concern has increased over the exploitation of mahogany in natural forest, as illustrated by the debate over the proposed inclusion of *S. macrophylla* on Appendix II of the Convention on International Trade in Endangered Species (CITES) (TFF 1994, Rodan and Campbell 1996). Such concern has encouraged increased research effort focusing on the ecology of the species in natural forest.

The potential of mahogany plantations as an alternative source of timber to natural forests has received rather less attention, largely as a result of the perceived shoot borer threat. Since Lamb's book, little information on mahogany plantation silviculture has been published and there have been few efforts to collate existing knowledge and present it in an accessible form. Yet evidence from many countries indicates that mahogany is a viable plantation species. The main aim of this book, therefore, is to produce a comprehensive account of mahogany silviculture in *plantations* by bringing together experiences and findings of foresters and researchers from mahogany-growing countries around the world.

An attempt has also been made to describe the silviculture of mahogany in *natural forest*. Unfortunately, evidence of positive silvicultural interventions is extremely limited. Despite the importance of managing natural forests sustainably, the lack of available information on appropriate silvicultural techniques for mahogany made it difficult to justify an independent treatment of natural forest silviculture. Those interested specifically in natural forests are referred to relevant sections in Chapters 2 and 13. There is clearly an urgent need for increased research effort on the silviculture of mahogany in natural forests, and it is our hope that the publication of this book may stimulate the collection of such information in the future.

The book is intended as a guide and reference document for anyone interested in growing mahogany. The layout resembles that of a standard silvicultural manual, with a logical progression through the stages of seed collection, plantation establishment, maintenance, prediction of growth and yield, protection and management of both natural stands and plantations under suitable silvicultural systems. In Chapter 11, options for control of shoot borer damage are discussed in detail. Although this chapter is found towards the end of the book, prospective growers may wish to refer to it at the outset, since strategies for controlling the shoot borer will affect many management decisions.

Most of the information that has been included is specific to the silviculture of mahogany; descriptions of standard silvicultural techniques have been kept to a minimum. Much information has been presented in tables in order to reduce lengthy discussion in the main text. Every effort has been made to report the findings, observations and beliefs of all sources as accurately as possible. The authors have tried to confine their own opinions to the conclusions at the end of the book. **All references to 'mahogany' can be assumed to indicate *Swietenia macrophylla* King (Meliaceae) only**. The scientific name is given where there is risk of confusion with other species in the same family.

The sources of information that were reviewed are extremely varied. Published information on mahogany in widely disseminated journals was relatively easy to obtain but often of limited relevance to silviculture. Less widely disseminated information, in the form of annual reports from forestry and research institutions, consultants' reports, university theses and in-country manuals, most of which are unpublished, proved to be considerably more useful. In order to obtain this information, research visits were made to Belize, Costa Rica, Fiji, Honduras, Malaysia, Mexico, the Philippines, Puerto Rico, St. Lucia, the Solomon Islands and Sri Lanka. Many different institutions, including forest departments, research institutes, research and development projects and private companies provided important documents and valuable assistance. Perhaps the most valuable sources of information were foresters and researchers themselves, many of whom had considerable experience in mahogany silviculture. Such inputs are cited as personal communications. Details of the individuals concerned are given at the end of the book, and their contributions are gratefully acknowledged.

To a large extent, the content reflects the availability of existing information. Some subjects, such as the economics of plantation or natural forest management, receive very little coverage because they have been poorly researched. It is also probable that some important sources of information have not been located. For example, although there is growing interest in the potential of mahogany silviculture in many South American countries, particularly Brazil, few of the findings generated in this region have been published in widely disseminated journals. More information is almost certainly available locally within such countries. It is hoped that perceived gaps in the book will stimulate the collection and presentation of relevant information in the future.

Mahogany is justifiably one of the best-known and most valuable tropical timbers currently traded internationally. The widespread development of plantations could make a major contribution towards meeting future demand for mahogany timber, and thereby help to reduce pressures on natural forest. It is the profound hope of the authors that this book will raise awareness of the potential of mahogany as a plantation species, and encourage the development of a plantation resource in appropriate areas.

Chapter Two

Description of the Species

2.1 Taxonomy and distribution of *Swietenia*

The name 'mahogany' is thought to be derived from a Yoruban word 'M'oganwo', literally meaning a 'group made up of many objects of great height', and was introduced to the Caribbean by Nigerian slaves (Lamb 1963). Lamb suggests that the term was originally used for trees of the genus *Khaya*, but that in the Caribbean 'mahogany' was used to describe a similar type of tree and timber, presumably *Swietenia* (see Keay 1996 for a historical account of the genus). However, Malone (1963) is sceptical of Lamb's etymology. If the word 'mahogany' was introduced as Lamb describes, Malone argues that it would have been generic, used for all 'tall trees' rather than just *Swietenia*.

Swietenia (Meliaceae) is generally considered to comprise three species: *S. macrophylla* King, *S. humilis* Zucc. and *S. mahagoni* (L.) Jacq. Taxonomic descriptions of these species are provided by Styles (in Pennington 1981, p. 390; see also Pennington and Styles 1975), who states that the three species are poorly defined biologically, partly because they hybridise freely. For example, naturally occurring hybrids of *S. humilis* x *S. macrophylla* occur in north-west Costa Rica (Holdridge and Poveda 1975, Whitmore 1983); variation in characteristics such as leaf morphology and fruit size is continuous in this area. Similarly, the hybrid between *S. macrophylla* and *S. mahagoni* arose spontaneously after the introduction of *S. macrophylla* to a number of Caribbean islands (Whitmore and Hinojosa 1977, Styles in Pennington 1981). The putative hybrid *S. humilis* x *S. mahagoni* has also been recorded (Whitmore and Hinojosa 1977).

The three taxa have been maintained as species on the basis of their different distributions, as well as morphological and ecological differences (Helgason *et al.* 1996). *S. mahagoni* is native to southern Florida (USA) and the greater Antilles, whereas *S. humilis* is distributed along the Pacific coastal region of southern Mexico and Central America, reaching its southern limit in Costa Rica. *S. macrophylla* is distributed more widely, ranging from southern Mexico throughout the Atlantic region of mesoamerica, reaching its southern limit in Brazil. *S. mahagoni* does not, therefore, occur sympatrically with either of the other two species within its native range. The distributional ranges of *S. macrophylla* and *S. humilis* overlap in a number of locations in Central America, which has led to uncertainty about the taxonomic identity of some populations, particularly those in Costa Rica.

Morphological differences between the three species are rather slight. Many of the vegetative characters listed by Styles (in Pennington 1981) display a considerable degree of variation, and a degree of overlap exists between the three species (Table 2.1). The usefulness of characters such as the micropubescence of the calyx margin has been called into question (Miller 1990). Preliminary molecular analyses of the genus *Swietenia* have also failed to differentiate the three species (Chalmers *et al.* 1994, Helgason *et al.* 1996). Whether or not *S. humilis* and *S. macrophylla* represent two separate species has been questioned; *S. humilis* could be viewed as an ecotype of *S. macrophylla,* and the genus *Swietenia* may even be considered to be monospecific (Helgason *et al.* 1996). However, differences in the ecological preferences of *S. macrophylla* and *S. humilis*, as well as their morphological characteristics, have led those taxonomists who have studied the genus intensively to maintain them as separate species (Styles in Pennington 1981).

Table 2.1. Taxonomic characters used to differentiate between *Swietenia* spp. (from Helgason *et al.* 1996; based on data provided by Styles in Pennington 1981)

Character	S. humilis	S. mahagoni	S. macrophylla
Seed colour	pale brown	chestnut	chestnut
Seed length (cm)	6-8(-9)	(2-)4-5(-6)	7.5-10(-12)
Calyx margin	ciliolate	not ciliolate	ciliolate
Lower leaflet attachment	sessile	petiolulate	petiolulate
Leaflet size (cm)	(4.5-)7-9(-14)	(2.5-)4-6(-8)	(7.5-)9-13(-18)
Leaf size (cm)	(12-)14-22(-30)	(10-)12-15(-28)	(14-)16-30(-40)

A lack of sufficient field collections has led to uncertainty about the precise distributions of the three species, and *S. macrophylla* in particular. The distribution maps presented by Lamb (1966) continue to be widely cited as the most reliable source (Fig. 2.1a,b). De Barros *et al.* (1992) presented revised maps for the estimated distribution of *S. macrophylla* in Brazil (Fig. 2.1c), where precise information is particularly lacking. The distributional area given by de Barros *et al.* (1992) is more equatorial than that of Lamb (1966), with a total area of 800,000 km^2. Within Brazil, *S. macrophylla* is most highly concentrated in an area of approximately 250,000 km^2 in southern Pará (de Barros *et al.* 1992). Further field research is clearly required to define the distribution and status of *S. macrophylla* more precisely.

2.2 General description of *S. macrophylla*

S. macrophylla is a large deciduous tree with an umbrella-shaped crown (Styles in Pennington 1981), frequently reaching heights of over 30 m and diameters at breast height (dbh) of over 1.5 m (Lamb 1966) (see Plate 2.1). Heights of 45-60 m and diameters of 2.5-3.5 m were not rare before such populations were extensively logged (de Irmay 1948). Mahogany is a canopy-emergent (Lamb 1966); in Bolivia, mahogany trees emerge from the canopy when they are 20-25 m tall and the tallest trees tower another 25 m over the canopy (Gullison 1995). The precise longevity of mahogany is unknown, although individual trees may apparently live for several centuries (Lamb 1966). The bole is cylindrical and is often buttressed (see Plate 2.2). The crown of young trees is narrow, but that of old trees is broad, dense and highly branched (Lamprecht

1989). Mahogany timber is prized particularly for its colour and workability; it is primarily valued for construction of high-value furniture and interior fittings (Palmer 1994). The colour and density of the wood vary markedly with geographic origin and growth environment (Lamprecht 1989).

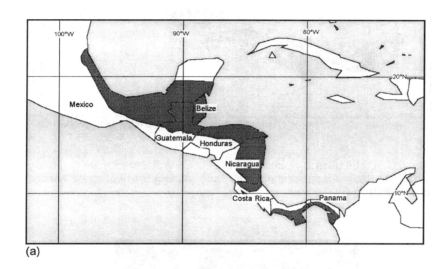
(a)

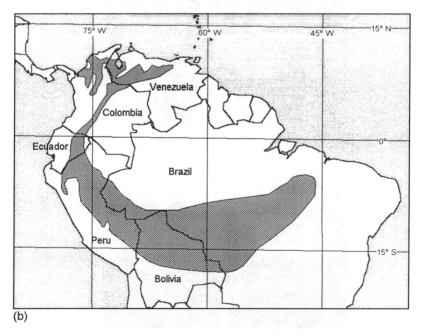
(b)

Caption overleaf.

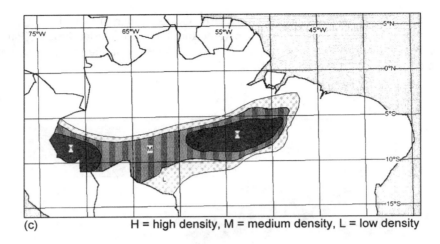

(c) H = high density, M = medium density, L = low density

Fig. 2.1. Distribution of mahogany in: (a) Central America (after Lamb 1966), (b) South America (after Lamb 1966) and (c) Brazil (after de Barros *et al.* 1992).

Plate 2.1. A mature mahogany tree showing characteristic shape. Las Caobas Ejido nursery, Quintana Roo, Mexico.

Plate 2.2. Buttresses on an old mahogany tree in natural forest. Rio Bravo Conservation and Management Area, Belize.

2.3 Ecological range

S. macrophylla is able to tolerate a very wide range of environmental conditions and is found naturally in both tropical dry and tropical wet forest types. Annual rainfall across its ecological range is typically 1000-2500 mm, reaching 3800 mm in Amazonian Ecuador and Peru (Lamb 1966, Whitmore 1983, Betancourt 1987, CITES 1997). According to Lamb (1966), mahogany reaches its 'optimum natural development' under tropical dry forest conditions, with an annual precipitation of 1000-2000 mm, a mean annual temperature of 24°C and a potential evapotranspiration ratio of 1-2. It has been recorded at altitudes of up to 1400 m in Peru and Bolivia.

Mahogany grows naturally on a very wide range of soil types, including those derived from sedimentary, igneous and metamorphic rocks, and those which are alluvial or volcanic in origin (Whitmore 1992). For example, in Quintana Roo, Mexico, mahogany is found on shallow soils developed on limestone, with a high organic content and a neutral pH (Negreros 1991). In Venezuela, the species is associated with light, well-drained, aerated soils, avoiding areas subject to prolonged inundation (Bascopé *et al.* 1957). In Brazil, mahogany is thought to be suited to well-drained upland alluvial plains (Lamb 1966). Despite the wide range of site conditions with which mahogany is associated in natural forests, the influence of environmental factors on growth have not been investigated in any detail.

In those forests within which it occurs, the population density of mahogany may vary widely (Gullison *et al.* 1996). In Brazil, Barros *et al.* (1992) considered densities of 1 mature tree ha^{-1} to represent the upper limit, with typical densities being much lower. Average density in unlogged Mexican forest is typically 1-2 mature trees ha^{-1} at most (Snook 1993). Densities of 4-8 mature trees ha^{-1} have been recorded in Venezuela

(Bascopé et al. 1957) and groups of 20-60 mature mahogany trees have been recorded in Bolivia, although even at the time of observation these were extremely rare (de Irmay 1948).

2.4 Reproductive biology

S. macrophylla is monoecious with unisexual flowers (Styles 1972) which are pollinated by insects. Bees and moths are believed to be the main pollen vectors (Styles and Khosla 1976), although there are also reports that thrips may act as pollinators (Patiño 1997). The flowers are 0.5-1.0 cm in length, and are borne in large, branched inflorescences including both male and female flowers. There have been reports of self-compatibility in *Swietenia* under experimental conditions (see Styles and Khosla 1976), but recent studies in Bolivia suggest that *S. macrophylla* may be obligately outbreeding, as are many other tropical tree species (Loveless and Gullison, cited in CITES 1997). No information is currently available on self-incompatibility mechanisms within this genus. After fertilisation, the fruit require 9-11 months to mature (Gullison 1995). An average of 55 (Gullison 1995) to 70 (Tito and Rodriguez 1988) seeds are found in each woody seed pod (see Plate 2.3).

Plate 2.3. Mahogany capsule showing the internal arrangement of winged seeds.

Although little information on reproductive phenology in *Swietenia* is available, flowering appears to be seasonal and simultaneous in most trees within a population (Styles 1972). The flowering period varies from region to region. In Quintana Roo, Mexico, mahogany starts flowering in March (Snook 1993), whereas in Puerto Rico the species generally flowers in May and June (Little *et al.* 1967). In the Chimanes Forest, Bolivia, flower and leaf production occur simultaneously in September at the onset of the rainy season (Gullison *et al.* 1996). Likewise, Burgos (1954) reported that flowering in Peru occurs between September and October. Leaf fall tends to occur at the onset of the dry season, leaving the mature capsules on the tree. The seeds are dispersed when wind breaks the columella to which the seeds are attached, after the capsule valves dry and fall off (Gullison *et al.* 1996).

Seed production from mahogany in both natural forests and plantations fluctuates considerably from year to year (Lobato 1969). The reasons for such fluctuations are not clearly understood, but may reflect variation in flowering phenology, or failure of pollination or fertilisation. In natural forest in Bolivia, Gullison *et al.* (1996) described a relationship between tree dbh and fecundity, defined in terms of capsule production per unit basal area. Maximum fecundity was recorded for trees of 90-130 cm dbh, perhaps reflecting a shift of allocation of photosynthates from growth to reproduction. The factors which determine the fecundity of individual trees are not clear. From observations of natural forest in Belize, Stevenson (1944b) notes that large quantities of viable seed are not usually produced by a tree of less than 60 cm dbh, but attributes this fact to a sub-dominant crown position. When the crown emerges from the natural forest canopy, which usually occurs when a mahogany tree is about 70 cm in dbh, large quantities of seed are produced and disseminated.

2.5 Regeneration ecology

2.5.1 Seed dispersal

Studies of seed dispersal are particularly relevant to those wishing to encourage mahogany stands to regenerate naturally. Although winged, seeds are relatively heavy and do not disperse far. In Bolivia, a median seed dispersal distance of 32-36 m and a maximum distance of over 80 m was estimated by Gullison *et al.* (1996). For mahogany trees in Belizean natural forest, seeds are spread over an area extending up to 90 m from an exploitable (70 cm diameter) tree (Wolffsohn 1961). In Mexico, seeds are generally dispersed a maximum distance from the tree which is approximately twice the tree's height (Santiago *et al.* 1992). The distance over which seeds can travel depends not only on the height of the tree, but the height and density of surrounding vegetation, the weight of individual seeds and the strength of wind at the time of release from the capsule.

Santiago *et al.* (1992) carried out a detailed study of mahogany seed dispersal in southern Mexico. An area of 8000 m^2 around a 30 m tall mahogany seed tree was cleared by bulldozer. A 100% enumeration was carried out, involving location of an estimated 84% of all seeds produced (16% were thought to have been blown out of the area). Seed densities ranged from 630 seeds ha^{-1} in the north-east quadrant to 17,390 seeds ha^{-1} in the north-west quadrant, due to the prevailing south-easterly breeze. Densities of 6490 seeds ha^{-1} and 5160 seeds ha^{-1} were found in the south-west and south-east quadrants respectively, highlighting the importance of northerly winds which are stronger and gustier than the prevailing wind but less frequent.

2.5.2 Conditions required for germination

The trigger for germination appears to be rainfall. In natural conditions, mahogany seed can survive short dry periods on the ground until rain stimulates germination (Tompsett 1994). The time taken for seeds to germinate varies from 2 to 16 weeks, depending on the amount of precipitation (Gullison 1995). However, once it has started, germination does not appear to be affected by variation in rainfall patterns in the wet season (Gerhardt 1994).

Gerhardt (1994) performed an in-depth study of above and below ground conditions that influence mahogany seedling establishment. The work was carried out in the Guanacaste Conservation Area in north-west Costa Rica. (There is some doubt as to the naturally occurring species in this area: Whitmore and Hinojosa (1977) found evidence of hybridisation between *S. macrophylla* and *S. humilis* and believe that populations of both species are present in the Guanacaste area, along with hybrid swarms.) The development of mahogany seedlings was monitored on abandoned pasture sites, successional deciduous forest and successional semi-evergreen forest, associated with different light intensities at ground level. The results indicated that germination rates of mahogany are unaffected by variation in light availability.

Germination may be prevented by predation by insects or birds. Wolffsohn (1961) reports the results of an experiment in the Chiquibul Forest Reserve of Belize. Sixty natural regeneration plots were marked out under 10 mature trees. Thirty of the plots were dusted with an insecticide (Aldrin) and 30 were left as a control. Aldrin was reapplied after every heavy rain shower. Naturally regenerated seedlings were counted after 6 months (height around 15 cm). On the 30 treated plots, 727 seedlings developed, but on untreated plots only 15 seedlings were recorded. Wolffsohn (1961) believed that practically all the seeds were eaten by weevils before the onset of the rains, half-way through the experiment. It was suggested that disturbance of habitats of predatory insects may account for the presence of regeneration along old logging roads.

In Mexico, Santiago *et al.* (1992) attributed low regeneration rates in some sample areas to seed predation, possibly by parrots (*Amazona xantholora*). Mahogany seeds are regarded as unpalatable to most mammals.

2.5.3 Conditions required for seedling establishment

S. macrophylla is generally considered to be a light-demanding species, requiring high light availability for seedling growth. Stevenson (1927a) noted the rapid response of suppressed seedlings to canopy opening, with leaders growing up to 60 cm in height over a 3 week period. The growth of young mahogany was thought to be directly proportional to the opening of canopy and undergrowth (Stevenson 1941). Gerhardt (1994) found that mahogany seedlings had a higher survival and growth rate in thinned than in unthinned treatments. Snook (1993) attributed the abundance of mahogany in log-landings and around the edges of seasonally flooded depressions in Quintana Roo to the high light availability on such sites (see Plate 2.4). These findings have been corroborated by experimental investigations of the ecophysiological responses of mahogany seedlings to variation in light availability (Ramos and Grace 1990).

Plate 2.4. Log-landing and harvesting crew in natural mahogany forest. Las Caobas Ejido, Quintana Roo, Mexico.

However, the failure of mahogany seedlings to grow under shady conditions does not imply that they are unable to survive. Seedlings appear to be able to tolerate a degree of shade, at least initially (Stevenson 1927a, Streets 1962, Lamprecht 1989). Stevenson (1936) suggests that the illumination required to ensure germination and establishment of mahogany seedlings may be less than that required for further growth. There is some evidence to support this hypothesis. In an experiment carried out in Quintana Roo, Mexico, 0.5 ha patches of natural forest were subject to different levels of overstorey removal (0, 8, 28, 45, 55% of basal area) and mahogany regeneration was assessed 4 years later (Negreros 1991, Garcia *et al.* 1993). The difference between treatments was not statistically different, indicating that light may not be an important determinant of seedling establishment. In some mahogany plantations, naturally regenerated seedlings have been observed to establish and persist, albeit with minimal height increment, under the dense canopies of mature trees for many years (Mayhew *et al.* in press, Kubuabola and Evo pers. comm.).

Establishment and survival is also determined by rainfall and soil moisture availability. On sites subject to seasonal drought, the shade of overstorey trees may aid survival of young seedlings by reducing drought effects. Gerhardt (1994) found that seedling survival in a semi-evergreen site (where dry season shade is 40-50%) was almost double that in a deciduous site (where dry season shade is 10-25%). The largest seedlings were most likely to survive. Gerhardt found evidence to suggest that reduced root competition by trenching around seedlings increased their chances of survival.

A survey of the natural forest improvement area in Silk Grass, Belize, led Stevenson (1927b, 1941, 1944b) to believe that observed distribution of naturally regenerated mahogany seedlings could be attributed to a combination of factors including soil, drainage and type of original vegetation. In general, the most suitable sites for mahogany were found in high forest on fine sandy loams overlying coarser micaceous loams with a fairly high clay fraction. Despite the presence of mature trees, natural regeneration was absent on steep well-drained slopes supporting dense populations of cohune palms (*Attalea cohune*) and lianes. This forest type is found largely in the south of Belize where precipitation is more than 2500 mm per annum. In contrast, forest characterised by the bayleaf palm (*Sabal excelsa*) contained much mahogany regeneration. This forest type was found largely in the north of Belize, where precipitation is less than 2000 mm per annum. In the north, cutting of the merchantable crop without additional silvicultural treatment appeared to stimulate sufficient replacement regeneration.

The absence of natural regeneration has also been attributed to the deep litter layer, the emerging radicle failing to penetrate to the mineral soil layers. Snook (1993) found that the highest density of mahogany seedlings was found on loosened, disturbed and exposed soil. Under such conditions (commonly found along roadsides), where mahogany seed trees are nearby, mahogany seedlings establish in large quantities. However, Wolffsohn (1961) demonstrated that the radicle could in fact penetrate the litter layer if the seed was protected from insect predation.

2.5.4 Natural regeneration and the role of disturbance

As mahogany seedlings require high light availability for successful establishment, the extent of regeneration in natural forests is strongly influenced by the incidence of disturbance to the forest canopy. A number of different investigations have indicated that distubances may be infrequent and their effect localised, leading to a unimodal diameter class distribution and an uneven or patchy distribution of regenerating stands. The disturbance mechanisms that have been proposed vary considerably.

Gullison *et al.* (1996) noticed a unimodal distribution of mahogany trees in research plots located in the Chimanes Forest, Bolivia. A larger inventory of 245 mahogany trees in the Chimanes Forest (prior to logging) produced a diameter class distribution with multiple peaks, indicating that the stands have undergone episodic regeneration (see Fig. 2.2). Gullison *et al.* (1996) explain the observed pattern by hydrological disturbance and suggest two mechanisms: gully erosion and log-jam-induced flooding.

Previous gully erosion appears to have increased the density of mahogany trees in natural forest on high terraces. In the area studied, erosion had ceased and little natural mahogany regeneration was found. However, the existence of mature mahogany trees along the edge of old gullies indicates that conditions for regeneration were once suitable, perhaps on disturbed upper slopes.

Log-jam-induced flooding results in deposition of sediment on the floor of low-lying forest. Although most species can survive flooding, mahogany appears better adapted to survive the high rates of sediment deposition (Gullison *et al.* 1996). (A related observation is made by Bramasto and Harahap (1989), who note that mahogany roots are able tolerate low oxygen concentrations for up to 70 days.) Where mahogany seed trees have survived, favourable conditions exist for the establishment of natural regeneration. The growth rate of mahogany seedlings was found to be significantly higher in areas of forest killed by river flooding and deposition than in undisturbed forest, as a result of increased light availability. Further studies are required to assess whether such processes affect mahogany populations in other areas of the Amazon basin.

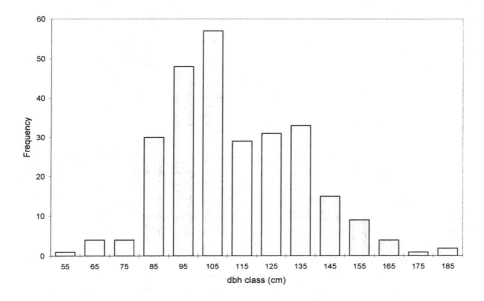

Fig. 2.2. Diameter distribution of 245 mahogany trees in unlogged natural forest, Chimanes, Bolivia (data replotted from Gullison *et al.* 1996).

A peak in abundance of mahogany trees in the 50-60 cm diameter class was recorded by Stevenson (1944b) in the forests of Belize. Profuse regeneration, established after wind damage to the forest canopy some time between 1865 and 1890, was thought to explain the diameter class distribution. Snook (1993) developed this disturbance theory for the adjacent forests of Quintana Roo, Mexico. Her research indicated that mahogany regenerated in essentially even-aged stands, the highest densities becoming established on areas which had previously been cleared, subjected to hurricanes or burnt. Particularly favourable conditions for regeneration were found to occur after hurricane damage followed by fire.

Hurricanes cause competing canopy trees to fall or break, creating gaps in the canopy ('top down' disturbance) and opportunities for seedling colonisation (Snook 1993). Mature trees of *S. macrophylla* appear to be relatively resistant to damage by hurricanes owing to heavy buttressing, strong but flexible wood, low density foliage and aerodynamic crowns. The fact that mahogany is more resistant to hurricanes than many other species is well known (Lamb 1966), but the explanations for this resistance vary, some suggesting the rapidity with which leaves are lost (Francis pers. comm.) or the deep tap root (Singh pers. comm.) as the key factor. Hurricanes also result in an increased volume of fallen timber, which may increase the subsequent risk of fire, particularly in the dry season.

Fire promotes mahogany regeneration by reducing the ground vegetation and providing suitable conditions for germination of mahogany seeds and growth of seedlings (i.e. 'bottom up' disturbance). Snook (1993) suggests that mature mahogany trees are protected from fire by their thick bark. However, there is some uncertainty over the resistance of mahogany to fire (see also Section 12.3). Mahogany does not sprout effectively after fire (Snook 1993) and the principal source of natural regeneration, therefore, is seed from surviving mature trees. Stevenson (1944b) described mahogany as 'fire-tender' and noted that large trees left standing in burnt areas were invariably killed. Trees standing in burnt clearings may not survive long enough to produce a crop of seed (Stevenson 1927c). There is evidence that mahogany regenerates well on burnt areas (Cree 1956, Frith 1959), but it is not clear in these studies whether seed-bearing trees themselves were located in the burnt area. Thus, although fires appear to assist mahogany regeneration, it is by no means certain that burnt areas are colonised by seed from mahogany trees which have survived burning.

Hurricane-damaged mahogany trees must recover their reproductive potential before the fire occurs, or competing vegetation will quickly occupy the site. It may be 2-3 years before mahogany trees bear fruit after a severe hurricane (Lamb 1946, 1948, Oliver 1992). Evidence indicates that forest fires associated with hurricanes are also delayed by a few years (Kinloch 1933, Lamb 1946, Lindo 1968), probably owing to the time taken for fallen timber to dry out. The delay gives mahogany sufficient time to recover and bear seed. However, the timing of such disturbance events, with respect to their impact on mahogany regeneration, is clearly critical.

Mahogany is also able to regenerate following some forms of forest disturbance caused by human activity. Irregular diameter curves of mahogany in forests of Yucatan, Mexico, are thought to be due to depopulation of the region and abandonment of agricultural land in the wake of the 'Guerra de Castas' of 1847 (Huguet and Verduzco in Lamprecht 1989). Many mahogany-rich stands in Mexico, Belize and Guatemala are known to be associated with earlier Mayan agricultural activity (Stevenson 1927c, 1944b, Snook 1993, in press). It is not uncommon to find mahogany growing on sites of Mayan ruins. Pollen analysis has indicated a marked increase in the abundance of Meliaceae (which may include *Cedrela odorata* as well as *Swietenia macrophylla*) following abandonment of Mayan agricultural land in Copán, Honduras, around the beginning of the 13th century (Abrams *et al.* 1996).

An association between mahogany and sites of previous human activity may also apply in South America. Pre-historic agricultural fields thought to date to between 500 and 2300 BC have been found throughout much of the Chimanes forest in Bolivia, where mahogany is now abundant (Gullison *et al.* 1996). It is conceivable that the current distribution of mahogany throughout both Central and South America may be partly a reflection of previous human activity. This intriguing possibility awaits further study of the region's archaeology and historical ecology before it can be fully evaluated.

2.5.5 Associated species

Because of its wide ecological range, *S. macrophylla* is found in many different forest types, ranging from swamp to sub-montane forest. Plant species associated with *S. macrophylla* also vary considerably throughout its geographical distribution. A description of the various plant associations in natural forest is outside the scope of this book; readers are referred to Lamb (1966) for a comprehensive account.

The best-known insect species associated with mahogany are shoot borers in the genus *Hypsipyla* (Lepidoptera, Pyralidae). *H. grandella* appears to be specific to *Swietenia* spp. and the closely related genus *Cedrela*. The larva of *Hypsipyla* spp. burrows into the terminal shoot of mahogany trees. The shoot may be killed, causing the tree to branch or produce multiple leaders. As a result, affected trees often have little or no timber value on reaching maturity. Considerable research has been devoted to mahogany shoot borers because of their importance as a forest pest (see Chapter 11).

Few data are available to evaluate the number of other organisms dependent on mahogany for survival. Numerous epiphytes colonise *S. macrophylla*, including orchids and bromeliads, but it is unknown whether any such species are specific to mahogany. Mahogany seeds are a food source for forest rodents (e.g. *Dasyprocta punctata* and *Cunicula paca*), parrots (*Amazonas* spp. and macaws) and a variety of insects. Mature mahogany trees are used as nesting sites by toucans (*Ramphastos* spp.) (CITES 1997).

Grazing mammals do not usually consume the foliage of young mahogany because of its bitter taste. In natural populations of mahogany seedlings in Bolivia, herbivory rates by all herbivores (including insects) were found to be very low compared with published studies of other tree species (Watkins and Gullison in press). Median rates were highest in gaps where 7.8% of leaf area was removed and lowest in the understorey, with 1.9% removed. There was no discernible relationship between the density of seedlings and herbivory rates, suggesting that the planting density of mahogany need not be reduced for fear of herbivory.

2.6 Summary

- *Swietenia macrophylla* has a wide geographical and ecological range, extending from southern Mexico to Bolivia and central Brazil. Within its natural range, stocking density of *S. macrophylla* is extremely variable and its distribution is patchy.
- Seed productivity of mahogany trees in natural forest fluctuates annually. Seeds are shed towards the end of the dry season and are rarely dispersed more than 100 m from the tree. Mahogany seeds can only retain their viability for a few months, not long enough for creation of a seed bank in the soil.
- Evidence indicates that germination is stimulated by rainfall rather than light availability. Germination may be prevented by seed predation, e.g. by insects or parrots.
- Mahogany seedlings do not establish easily in undisturbed natural forest. Many factors including light, climate, litter layer, soil and drainage affect the survival of germinating seedlings. Once established, mahogany seedlings may survive under a closed canopy for several years. However, mahogany is a light-demanding species and for further growth to take place, increased illumination by canopy opening is required.
- Many natural mahogany stands show a unimodal diameter class distribution. This implies that favourable conditions for establishment and growth occur only sporadically.
- In the forests of the Amazon basin, the distribution of mahogany-rich stands has been linked to hydrological processes. Higher than average densities of mahogany are found in areas which have been subject to gully erosion and log-jam-induced flooding.
- In the natural forests of Belize and Mexico, the presence of mahogany-rich stands has been explained by a combination of hurricanes and fire damage. Mahogany is able to survive hurricane damage better than most species and may display a degree of fire resistance.

- The abandonment of agricultural fields in Central America after the collapse of the Mayan empire may help to explain the current distribution of mahogany-rich forest. Likewise, in parts of the Amazon basin, human disturbance through shifting cultivation may have encouraged mahogany regeneration.

Chapter Three
Mahogany as a Plantation Species

In 1980 there were an estimated 55,200 ha of mahogany plantations around the world (Pandey 1983). There is insufficient information to produce a precise global estimate for 1998, but just over 200,000 ha have been reported (see Table 3.1). Included in this figure are areas of natural forest that have been line planted with mahogany.

The largest mahogany plantations are located in South and South-East Asia and the Pacific regions. A significant proportion of the total area, most notably in Indonesia and the Philippines, has been designated for protection of slopes and water catchments and may not be productive. In addition, mahogany has been widely used for avenue planting (see Plate 3.1). Attempts to introduce mahogany to tropical African countries were largely unsuccessful due to the severity of shoot borer attack on young plants. There is little evidence of extensive planting across the natural range of the species (Central and South America).

The majority of mahogany plantations about which information has been collected are state-owned, but there are a few exceptions. About 30% of the plantations in the Solomon Islands are privately owned (Kolombangara Forest Products Ltd.). Ownership of many Filipino plantations will pass into the hands of local communities under the community-based forest management scheme. The Fijian Forestry Department is currently (in 1998) considering the possibility of privatising its mahogany plantations (Singh pers. comm.).

3.1 Origin of seed used to establish mahogany plantations

Dates of introduction and sources of mahogany seed have been recorded in a number of countries (see Table 3.2). Unfortunately, the records are often incomplete, and in some cases it is likely that earlier, undocumented introductions were made. For example, the earliest recorded introduction of mahogany to any country is to Indonesia in 1870 (see Table 3.2). However, the documented source of this seed was India which apparently did not receive any seed until 1872, when a consignment (mistakenly labelled *Swietenia mahagoni*) was received by the Royal Botanic Gardens in Calcutta (Perera 1955). If the Indonesian report is correct, mahogany seeds must have been introduced to India earlier than stated. It is interesting to note that the shipping of *Swietenia mahagoni* seeds began much earlier. The species was introduced to Sri Lanka, almost certainly via India, and planted as an avenue tree in Jaffna as early as 1840 (Vincent 1884).

Plate 3.1. An avenue of mahogany trees. Salinas Reforestation Project, Luzon, Philippines.

The origin of the mahogany seed which was shipped to South and South-East Asia in the late 19th century is in some doubt. India is widely quoted as the source, but origins of Indian seed are unknown. At the time, introductions of mahogany were probably the result of trade between outposts of the British empire. Belize (then British Honduras) or parts of the Mosquito Coast (now Honduras and Nicaragua), the only regions colonised by the British in the natural range of *S. macrophylla*, are the most likely countries of origin. Belize was the origin of some of the first mahogany seed to reach Sri Lanka (see Table 3.2). The earliest recorded introduction to South-East Asia of mahogany seed which was definitely of a non-Belizean origin occurred some time before 1938, from Guatemala to Indonesia. By the beginning of the 20th century, plantations in India, Indonesia and Sri Lanka were producing sufficient seed for export to other countries in the region, notably Malaysia and the Philippines.

The sources of introductions to the Pacific region from the end of the Second World War until the 1990s are more varied (see Table 3.2). In the 1960s, growing interest in mahogany in Fiji led to fresh imports of seed from plantations in Sri Lanka and the Philippines to supplement seed from small existing blocks of trees grown from Belizean seed. More seed was received by the Fijian Forestry Department from Trinidad and Singapore (origins unknown). The Solomon Islands also received seed from Trinidad (origin unknown) and Honduras. By 1990, Sandiford (1990) reports a total of 41 recorded accessions of mahogany from overseas to the Solomons, the majority coming from Fiji and of unidentified provenance. Seed from Fijian plantations was exported to various other Pacific islands (and to Malaysia).

With the exception of Puerto Rico, the introduction of mahogany to Caribbean countries is poorly documented (see Table 3.2). Private planters probably introduced the species before Forest Departments, as in Jamaica (Streets 1962).

Table 3.1. Countries with documented mahogany plantations

Country	Area (ha)	Function	Description	Source
Antigua	-	-	nda	Anon. (1944)
Australia	-	Trials	Trial plantation in Queensland was found to be reasonably successful but uneconomic	Keys & Nicholson (1982)
Bangladesh	-	Avenues	Mahogany is widely planted along roadsides, but not commonly found in plantations	TFF (1994), Sarshar (pers. comm.)
Belize	100	Production	Relatively large areas (of which 630 ha are documented) were established by the Forest Department from 1930 to 1960. All plantations have been neglected since 1960 and most have been felled, in some cases by milpa cultivators unaware of their value	Forest Department Annual Reports (1945-63), Ennion (1996)
Brazil	-	-	nda	TFF (1994)
Costa Rica	-	Trials + production	Mahogany has often been established by private landowners in coffee plantations, but is not favoured by the Forest Department. Some small-scale trials have been set up	Somarriba (pers. comm.)
Cuba	-	-	nda	Geigel (1977)
Dominican R.	-	-	nda	Weaver (1995)
Ecuador	-	-	nda	Flinta (1960)
Fiji	42,000	Production	Plantations establishment began in 1962 and large-scale production is now imminent. Of the total mahogany plantation area, 50% is located in the Southern Division of the main island, Viti Levu. Plantations have also been established on Vanua Levu, but are still young. Mahogany has been the only plantation species used in Fiji since 1991	DoF (1993), Seroma (1995), Singh (pers. comm.)
Grenada	-	Trials + production	Plantations were established after hurricane Janet in 1955 by both the forestry department and private estate owners	Frederick (1994)
Guadeloupe	4200	Production	Most plantations have been established since 1948	Soubieux (1983)
Guam	-	-	nda	Krohn (1981)
Guyana	-	Avenues	Mahogany has been used for avenue planting in Georgetown; there are no plantations	Turnbull (pers. comm.)
Haiti	-	-	nda	Timyan (pers. comm.)
Hawaii	-	Trials	Mahogany is reportedly well adapted to conditions	Nelson & Schubert (1976)
Honduras	150	Production	All that remains of the United Fruit Company's mahogany plantations can be found in Lancetilla Gardens, Tela. These were planted in 1946. Originally about 1830 ha had been planted, the majority (1520 ha) between 1948 and 1950. Recently there has been plantation of gaps in natural forest under the Proyecto de Bosque Latifoliado	Chable (1967), Hazlett & Montesinos (1980), Meza (pers. comm.)
India	-	Trials	The first plantations were established in Chittagong (now Bangladesh) and in southern India at low elevations. By 1962 plantations were reported only from Madras state	Troup (1932), Streets (1962)

Table 3.1 (continued). Countries with documented mahogany plantations

Country	Area (ha)	Function	Description	Source
Indonesia	116,282	Production + protection (at least 33%)	Initial trials were a success and 'large' (unspecified) areas were established at the end of the 19th century. Mahogany was planted on sites not sufficiently fertile for teak and in gullies to reduce soil erosion. Since 1987 mahogany has been widely planted in Java by the Forestry Board (Perum Perhutani) and so the majority of existing plantations are immature. Plantations are also found in Sumatra (2500 ha), West Timor and Sulawesi	Noltee (1926), Bramasto & Harahap (1989), Fattah (1992), Fox (pers. comm.)
Jamaica	-	-	nda	Wadsworth (1960)
Malaysia	-	Trials	There has been little planting in Peninsular Malaysia or Borneo. At Luasong, Sabah, 200 ha (now owned by the Innoprise Corporation) was established, but without success	Kotulai (pers. comm.)
Martinique	1479	Production	Mahogany stands are almost pure (94% of basal area). Today 28% are more than 32 years old. Mahogany has been planted in Martinique since 1924 and is currently in its second rotation. There are small areas (100 ha) of private plantation	Poupon (1982), Tiller (1995), Chabod (pers. comm.)
Mauritius	-	Trials	Mahogany plantations were initially unsuccessful, but later used successfully for reforestation at low elevations	Troup (1932), Perera (1955)
Mexico	-	Production	Many small blocks have been planted on private farms. Enrichment planting of communally owned forest has been carried out along skid trails and on log-landings	Lamb (1966), Negreros (in press), Tipper (pers. comm.)
Montserrat	-	-	nda	Anon. (1944)
Myanmar	-	-	nda	TFF (1994)
Nicaragua	-	-	nda	Lamb (1966)
Nigeria	-	Trials	Introduction has been largely unsuccessful	Streets (1962), DFR (1963), Akanbi (1973), Lamb (1966)
Panama	-	-	nda	
Peru	-	Production	Small plantations were established by lumber mills as early as 1946. More recently, a Japanese project established 700 ha of mahogany in Peru, though with little success owing to the shoot borer	Hoy (1946), Ikeda et al. (in press)
Philippines	~25,000	Protection + production	Mahogany is one of the five most planted species in the national reforestation programme. Around 50% of the current (estimated) plantation area was established between 1988 and 1992: 3940 ha in Zamboanga, 2160 ha in Western Visayas, 1780 ha Southern Tagolog and 1630 in Ilocos. Mature stands (perhaps around 1000 ha) serve as seedlots. There are also areas, possibly several thousand hectares (not included in the total figure), of private plantation	DENR (1992), Lapis (pers. comm.), Patricio (pers. comm.)

Country	Area (ha)	Purpose	Comments	References
Puerto Rico	650	Protection + trials	Around 270 ha of *Swietenia macrophylla* were established in the Luquillo Experimental Forest in 1936-39. Hybrid mahogany (*S. macrophylla* x *mahogani*) was used when planting recommenced in 1963; most has been line planted. Early *S. macrophylla* plantations in limestone sinkholes at Rio Abajo showed rapid growth and impressive form, as do provenance trials on similar sites in Guajataca	Marrero (1950), Bauer & Gillespie (1990), Francis (pers. comm.), Wadsworth (pers. comm.),
St. Lucia	101	Protection + production	Mahogany has been established for watershed protection; stands not damaged by hurricanes contain trees of good form, particularly in mixed *Hibiscus elatus* stands	Andrew (1994), Bobb (pers. comm.)
Seychelles	-	Trials	*nda*	Anon. (1955b)
Sierra Leone	-	Trials	Introduction has been largely unsuccessful	Streets (1962)
Solomon Islands	3065	Trials + production	Found on a number of islands including Guadalcanal, Munda and Kolombangara. There has been recent interest in planting on some of the eastern-most islands (such as Santa Cruz), which are free of the shoot borer	Chaplin (1993), Speed (pers. comm.)
Sri Lanka	4500	Production	Most mahogany plantations are located on the eastern side of the island (the Intermediate Zone). Many of the plantations date from the 1930s and 1940s, when mahogany was established in young *Artocarpus integrifolia* plantations. Stands are uneven-aged owing to selective felling practices and natural regeneration. Most are now dominated by mahogany. Line planting in wetter parts has been unsuccessful	Sandom & Thayaparan (1995); Sandom (pers. comm.)
Surinam	-	-	*nda*	Negreros (in press)
Taiwan	-	Trials	Plantations established in the early 1950s have been reported at the Chung-pu Branch Station of the Taiwan Forestry Research Institute	Lee (1968)
Thailand	623	Trials	Plantation began in 1964, although 98% of the current area was established after 1985.	Boontawee (1996)
Tonga	-	Trials	Recently established plantations are disappointing	Oliver (1992)
Trinidad and Tobago	-	Production	Most planting has been carried out on cocoa and other estates, usually along roadsides or estate boundaries. There are few examples of well-managed mahogany plantations	Lamb (1955), Pawsey (1970)
Uganda	-	Trials	Trials took place in the 1930s and 1950s. Rapid early growth was observed, but the shoot borer caused serious damage and the introduction was deemed unsuccessful	Streets (1962), Kriek (1970)
Vanuatu	-	Trials	Trials have been carried out successfully, but large scale planting has yet to take place	Leslie (1994)
Venezuela	-	-	*nda*	Weaver (1995)
W. Samoa	2295	Production	Plantations have been established recently and are immature	Iakopo (pers. comm.)

nda = no details available

Table 3.2. Dates of introduction and origin of mahogany seeds

Country (region)	First record	Natural seed sources (intro. date)	Plantation seed sources (intro. date)	Source
South and South-East Asia				
India	1872	Belize? (1872)	nda	Troup (1932)
Indonesia (Java)	1870	Guatemala (pre-1938)	India (1870)	Noltee (1926), Coster (1938)
Malaysia (Peninsular)	1886	nda	Sri Lanka (1886), India (1899), Sri Lanka (1928)	Streets (1962), Hepburn (1969), Butt & Chiew (1982)
(Sabah)	1953	nda	Fiji (n.d.), Costa Rica (n.d.)	Ata & Ibrahim (1984), Chai & Kendawang (1984), Appanah & Weinland (1993), Kotulai (pers. comm.)
(Sarawak)	1926-8	nda	nda	
Philippines	1907	Peru? (n.d.), Mexico (1958)	India (1913), Trinidad (pre 1955), Jamaica (1958)	Ponce (1933), Chinte (1952), Anon. (1955a), Patricio (pers. comm.),
Sri Lanka	1889	Belize (1890)	India? (1889), India (1921)	Lushington (1921), McNeil (1935), Perera (1955), Tisseverasinghe & Satchithananthan (1957)
Thailand	1909	nda	nda	Boontawee (1996)
Pacific				
Fiji	1898	Belize (1911 and subsequently)	Singapore (post 1940), Trinidad (post 1940), Sri Lanka (pre 1955 and 1960s), Philippines (1960s)	Cottle (no date), Perera (1955), Streets (1962), Wright (1968), Singh (pers. comm.)
Solomon Islands	nda	nda	Fiji (n.d.), Singapore (n.d.), Philippines (1957), Honduras (1958), Trinidad (1958), Philippines (1977), Sri Lanka (1977), Holland? (1981)	Forest Division (1987, 1990), Sandiford (1990)
Tonga	nda	nda	Fiji (n.d.)	Oliver (1992)
Vanuatu	nda	nda	Fiji (n.d.)	Neil (1986)
Western Samoa	nda	nda	Philippines (n.d.)	Oliver (1992)
Caribbean				
Guadeloupe	1942	nda	Martinique (n.d.)	Soubieux (1983)
Jamaica	nda	Belize (n.d.)	nda	Streets (1962)
Martinique	1904	nda	nda	Tillier (1995)
Puerto Rico and Virgin Islands	1904	Belize (1904), Venezuela (1918), Panama (pre 1942), Peru (n.d.)	nda	Marrero (1942), Lamb (1960)
Trinidad	1906	nda	nda	Lamb (1955)
Central America				
Belize	1962	Mexico (1962-63)	Virgin Islands (1962)	Frith (1963, 1964)
Honduras	nda	Nicaragua (1940s), Guatemala (1940s)	nda	Chable (1967)

? = source uncertain   n.d. = no date   nda = no details available

Mahogany seed was also introduced to countries that already contained natural mahogany populations (see Table 3.2). Imports were often a response to a shortfall of seed before a new planting programme. For example, seeds were imported for planting in Columbia River Forest Reserve, Belize, because of seed shortage after the hurricane of 1961. Given the size of the plantation resource in Belize, it is unlikely that plantations grown from imported seed still survive. In Honduras, plantation establishment was carried out by the United Fruit Company on company-owned lands allocated to the growth of saw timber (Chable 1967). Mahogany seed was initially collected locally in Lancetilla valley, but additional supplies were later brought from Nicaragua and Guatemala. Some of the latter was gathered from the Pacific coast and may have been *S. humilis*. All seeds were collected behind logging operations in natural forests and the phenotypes were therefore of high quality. It is possible that it was seed from these plantations, rather than natural forest, that was sent to Fiji in the 1950s. The only remaining United Fruit Company plantations in Honduras are in Lancetilla Gardens and their provenance is unknown.

The current distribution of mahogany as an exotic is extremely wide, but available evidence suggests that the genetic base of the plantations in most countries is very narrow, with a large proportion of Belizean origin. There are fears in the Philippines that further genetic erosion may have occurred during the process of domestication (Gillies pers. comm.). A narrow genetic base may have been responsible for the disappointing results of some recent plantings, e.g. seed from Fiji, planted in Tonga (Oliver 1992). Where mahogany is planted as an exotic, it may be necessary to introduce new genotypes into existing plantation programmes.

3.2 Attempts to improve quality of planting stock

3.2.1 Provenance trials and progeny tests

Very few data are available from genetic tests of mahogany, as few tests have been established (Newton *et al.* 1993b, 1996). Collection of *Swietenia macrophylla* seed was carried out from 18 different areas in Central America ranging from Mexico to Panama during 1964-65 and provenance trials were subsequently established in a number different climatic zones in Puerto Rico and St. Croix (Geary 1969, Geary *et al.* 1973). After 8 years, the northern-most seed source of *S. macrophylla*, found in Mexico near Tampico on the Caribbean coast, had lowest survival and growth rates, but the seeds were collected from an area where dysgenic selection had apparently occurred (Geary *et al.* 1973). On wet planting sites, the fastest growing trees originated from areas with the shortest dry season (north-east and south-east Nicaragua) and the slowest growing trees from areas with the longest dry season. On dry planting sites, shoot borer attack tended to restrict height growth by killing new shoots, making the magnitude of height differences produced by provenance difficult to evaluate. The trial plots were remeasured at the age of 21 years and the initial conclusions were found still to be valid (Glogiewicz 1986).

More recently, two genetic tests of *S. macrophylla* were established in Costa Rica and Trinidad (in 1991 and 1990 respectively), by the Centro Agronómico de Investigación y Enseñanza (CATIE), Costa Rica, and the Trinidadian Forest Service respectively. The trials were assessed for height growth and incidence of pest attack 30-33 months after establishment. In the former trial, the maximum family mean was 192% of the minimum value, giving an indication of the extent of genetic variation recorded. The effect of genetic origin (either provenance or family) was statistically highly significant in both of the trials. Individual-tree narrow-sense heritabilities of height growth in the Costa Rican and Trinidadian trials were 0.38 ± 0.12 and 0.11 ± 0.06 respectively. Although these results are preliminary in nature, based on tests with only a limited range of material and after fewer than 4 years' growth, they indicate clearly that significant genetic variation exists both within and between different mahogany populations in terms of height growth (Newton et al. 1996).

One of the purposes of these trials was to screen for resistance against attack by shoot borers (*Hypsipyla grandella*). The response of *S. macrophylla* to shoot borer attack was assessed by scoring each tree in the genetic tests for the number of damage sites (branching or forking points). In the progeny tests established in Costa Rica and Trinidad, family mean number of attack sites varied between 2.2-4.7 and 2.7-4.4 respectively. The most pronounced genetic variation was recorded in the trial at CATIE, where the maximum family mean was 217% of the minimum value. The effect of genetic origin (either provenance or family) was statistically significant in both of the trials. Individual-tree narrow-sense heritabilities of number of forking points in the Costa Rican and Trinidadian trials were 0.56 ± 0.15 and 0.42 ± 0.12 respectively (Newton et al. 1996). The results from these trials indicate that significant genetic variation exists within *S. macrophylla* in terms of susceptibility to pest attack. This suggests that there may be scope for the development of pest-tolerant planting stock for use in reforestation, through a programme of genetic improvement (Grijpma 1976, Newton et al. 1993a,b). Long-term trials, testing a wider range of material, are required to develop this approach.

Patiño (1997) describes a provenance/progeny test established in 1988 in Campeche, southern Mexico, in which 36 progenies of three provenances of *S. macrophylla* were included. After 6 years, heritabilities (h^2) for height growth ranged from 0.038 to 0.265, depending on provenance. Provenance trials were also carried out in Vanuatu (see Table 3.3). Survival was low owing to cyclone damage and infection with brown root rot (*Phellinus noxius*), although the plots were unaffected by the shoot borer. On islands at risk from hurricanes, survival is an important consideration and Leslie (1994) recommended the use of the Honduras (Lancetilla) provenance. The Puerto Rico (Botanic Gardens) provenance was not considered as suitable because of its probable narrow genetic base. The high survival of the Vanuatu and Fiji provenances may well reflect the development of hurricane-resistant land races (Leslie 1994), as with *Pinus caribaea*.

At Instituto Nacional de Investigaciones Forestales y Agropecuarias (INIFAP) in Quintana Roo, *S. macrophylla* was collected from 5 local families and a trial has been set up (Parraguirre pers. comm.). No mensuration data are currently available.

3.2.2 Genetic improvement

Provenance and progeny tests are generally central to any genetic improvement programme with forest trees. The paucity of such tests with mahogany is indicative of the general lack of genetic improvement programmes with the species. Attempts to improve the quality of mahogany trees have been made in Fiji and the Philippines (Malvar pers. comm., Singh pers. comm.). In Fiji, a clonal seed orchard has been established as part of a mahogany tree improvement programme. Shoots of plus trees have been grafted on to root stocks with an 88% grafting success rate (SRD 1993). Similarly in Campeche,

Mexico, a seed orchard was established with 25 selected clones in 1990 (Patiño 1997). The orchard has yet to produce seed.

Table 3.3. Results of *S. macrophylla* provenance trials in Vanuatu after almost 6 years (Leslie 1994)

Seedlot	Height (m)	dbh (cm)	Survival (%)
Guarabo, Puerto Rico*	6.50	10.8	6
Guanica Forest, Puerto Rico*	13.75	18.0	31
Rio Piedras, UPR Botanic Gardens, Puerto Rico	15.86	19.0	41
Lancetilla, Honduras	14.78	18.5	42
Bacaramanga, Colombia	16.16	19.7	31
La Venta, Honduras	7.73	9.3	17
Colo-i-Suva, Fiji	16.08	14.6	52
Tagabe, Efate, Vanuatu	13.86	16.1	58

* *S. macrophylla* x *S. mahagoni* hybrids

Grafting with mahogany has been studied in detail by Zabala (1977). Ordinary root stocks were used, with scions from plus trees. Root stocks were 1.3 years old at the time of grafting, with a root collar diameter averaging 1 cm. Scions were taken from the upper portion of plus tree crowns, cut back to 16-24 cm, with all leaves except the young shoots removed. Grafting was found to be most successful when the scions were taken at the time of flushing, giving 59% survival. However, when scions were collected from plus trees 3 months later, survival dropped to 20.6%.

Under seed orchard conditions, seed production does not begin for 8 years (Kubuabola pers. comm.), and it will be many more years before large quantities are produced. In the mean time, efforts are being made in countries with a sufficient mahogany resource (e.g. Sri Lanka, Philippines and Fiji) to increase the proportion of seed collected from selected 'plus' trees (Patiño 1997).

These different initiatives indicate that many of the key elements of a genetic improvement programme with mahogany exist in different places. What is lacking is an integrated and intensive approach in any given area. The preliminary results available from provenance and progeny tests (see Section 3.2.1.) indicate that significant genetic variation exists within mahogany. Such variation could be exploited by selection, perhaps involving the establishment of seedling seed orchards to provide a source of genetically improved seed. Similarly, the incorporation of phenotypically superior 'plus' trees in clonal seed orchards may ultimately provide sources of improved planting material, although to be effective such an approach would have to be accompanied by progeny testing to identify genetically superior trees within the orchard (Zobel and Talbert 1984). The successful development of vegetative propagation techniques with mahogany (see Section 5.2) also raises the possibility of genetic improvement using clonal approaches, which may be of particular value in capturing pest-resistant genotypes for use in reforestation (Newton *et al.* 1994).

3.3 Summary

- The total recorded mahogany plantation area around the world is approximately 200,000 ha, although this figure excludes the majority of private plantations.
- Mahogany has been widely planted outside its natural range, in the Caribbean, South Asian, South-East Asian and Pacific regions. In each region, examples of both successes and failures can be found. Trials were established in many African countries, but generally proved unsuccessful.
- There are fewer mahogany plantations within its natural range. Until recently, availability of naturally grown timber, combined with problems caused by the shoot borer, reduced the incentive to establish plantations. However, plantations have been successfully established in Belize, Honduras and parts of Brazil.
- The largest planted areas are currently found in Indonesia (116,000 ha), Fiji (42,000 ha) and the Philippines (an estimated 25,000 ha). Planting programmes in Fiji, Indonesia, the Philippines and Western Samoa have been responsible for a rapid increase in plantation area in recent years. Mature plantations can be found in Fiji, Guadeloupe, Indonesia, Martinique, the Philippines and Sri Lanka.
- It is impossible to document seed origins for many individual plantations. There is little evidence to indicate that South American provenances have been widely exported or planted. Outside the natural range of mahogany, many of the world's plantations appear to have been established using seed of Belizean origin. It seems likely that the genetic base of plantation mahogany is extremely narrow. Plantation quality may have suffered as a result, and there is a need to broaden the genetic base of many plantations through new introductions.
- Few attempts have been made at genetic improvement of mahogany. Preliminary results from provenance tests indicate that significant variation exists in growth and form characteristics between provenances. There is therefore considerable potential for improving the quality of future plantations through careful selection and breeding.

Chapter Four

Seed Production

Mahogany starts to produce viable seed from a young age, at least in plantations (see Table 4.1). As with most plantation species, crown development and seed production is greatly affected by tree spacing and hence thinning regimes. However, early observations in Fiji (see Table 4.1) indicate that well-spaced trees were sometimes slow to set seed. One possible explanation is that the trees in question were damaged by strong winds; mahogany seed production after a Pacific cyclone may cease for 3 years (Oliver 1992).

Table 4.1. Age of planted mahogany trees at first seeding

Age (yrs)	Description of mahogany stand	Country	Source
8	Clonal seed orchard, 9 x 9 m spacing	Fiji	Kubuabola (pers. comm.)
8-9	Isolated individuals	Peru	Tito & Rodriguez (1988)
8-10	Spacing initially 2 x 2 m	Philippines	Ponce (1933), Chinte (1952)
10+	10 x 3 m spacing in enrichment lines	Solomon Islands	Chaplin (1993)
12	'Favourable conditions'	Central America	Lamb (1966)
12	Spacing initially 2 x 4 m	Mexico	Parraguirre (pers. comm.)
12-13	nda	Indonesia	Noltee (1926)
12-13	nda	India	Troup (1932)
<15	Spacing initially 3 x 3 m	Trinidad	Marshall (1939)
15	Planted under young *Artocarpus* sp.	Sri Lanka	Perera (1955)
18-20	nda	Fiji	Streets (1962)
20	Recently thinned	Sri Lanka	Streets (1962)
23	Basal area 25 m^2 at time of seeding	Puerto Rico	Trop. For. Res. Centre (1959)
30	Small stands and avenues	Fiji	Busby (1967)

nda = no details available

4.1 Seed collection

The inaccessibility of remaining natural mahogany stands and the relatively small areas of seed-producing plantations have greatly limited the supply of mahogany seed. Perhaps due to the cost of imports and limited seed longevity (see Section 4.3), forest managers have tended to collect seed from local sources. The first large-scale mahogany planting phase in Fiji (in the 1960s) relied on local seed collections from only a few stands and avenues; every tree was utilised (Busby 1967). During the establishment of plantations in Belize, seed shortage was a recurring problem and seriously affected both planting densities and plantation area targets (Lamb 1946, 1948, Cree 1956, Frith 1960). Recurrent seed shortage in these and other countries has made it impossible to confine seed collections from phenotypically superior 'plus' trees. Plantation quality has almost certainly suffered as a result.

Despite the growing area of mature plantations, seed from plus trees is still in short supply. For example, only 1.4% of seed from Fiji (which has the largest area of seed-producing mahogany plantations) is collected from plus trees; 33.7% is collected from selected seed stands and 64.9% from unselected trees (SRD 1995).

Different countries use different techniques for seed collection. In St. Lucia, tree climbers use spikes to climb mahogany trees and pull off near-mature capsules (Bodley pers. comm.). Tree climbers in Puerto Rico use bamboo poles to dislodge the seed capsules, which are collected just before dehiscence (Bauer 1987b). In the Philippines capsules are gathered by hand picking or pole picking before they are open (ERDB 1996). There are a number of difficulties associated with seed collection. Recent climbing safety regulations in Puerto Rico make collection difficult or prohibitively expensive (Francis pers. comm.). In Sri Lanka plus trees may be 35 m tall and forest officers have found that most local tree climbers are unwilling to climb so high (Tilakeratne pers. comm.).

The timing of seed collection is important. Seeds collected from ripe, harvested fruit have a greater germinative capacity than seeds from fallen fruits (Streets 1962, Anon. 1973). According to Busby (1967), a sample of seeds should be collected when capsules start to change colour from grey-green to brown, to determine if picking should commence. Fruit is ready for picking from an individual tree when:

- the capsules begin to show signs of contraction along the lines of dehiscence;
- the internal septa have begun to change from a cream to a pale brown colour;
- all the seeds are a uniform dark brown.

Immature seed has a low initial germination viability and demonstrates a rapid decline in viability during storage.

Chable (1967) identified the following characteristics of healthy seeds:

- the seed coat should be brittle, free of insect holes, firm to the touch (not spongy), with a burnished copper lustre;
- the seed itself should be pale creamy white at the centre, shading to a barely discernible pale green near the seed coat.

Parraguirre and Cetz (1989) carried out an experiment in Mexico to determine the best time to collect mahogany seeds in order to achieve high rates of germination. Work was carried out from the beginning of the fruiting season (mid-January), though did not continue through to the end. Capsules were taken from the trees and exposed to the sun until they opened. Seeds were extracted and sown immediately. The results showed that it was not until 1.5 months after fruiting began that germination reached more than 60% (double the average germination of the previous month). The increase in germination corresponded to a reduction in moisture content of the seeds from 45% to below 30% over the same period. Germination rates may have been higher at the end of the season (April/May), but the experiment was terminated at the end of March. The results do not preclude the possibility that higher germination rates could be obtained earlier by drying the seeds artificially.

In the Solomons, the peak seed collection period lasts around 2 months (Anon. 1987a). However, it is sometimes possible to collect seed over a 6 month period.

4.2 Seed preparation

Seed should be extracted from the capsules as soon as possible after picking, since rot may set in after only 2-3 days (Busby 1967). Unripe capsules may need to be dried out first. Early practice in Fiji was to place the capsules on a rack over electric lamps at a temperature of 38°C for 36-48 hours to encourage them to open. Alternatively, capsules can be dried in the sun. Estimates of the required drying period for the capsule vary from 1 day (ERDB 1996), to 2-4 days (Bauer 1987b, Bodley pers. comm.) to 5 days (Tito and Rodriguez 1988). The length of time required will depend on the ripeness of the capsule and ambient temperature and humidity. The capsules may be broken open by a sharp blow and the seeds are extracted and separated by hand (Busby 1967).

Many authors describe the removal of the seed wing (Busby 1967, Bauer 1987b, Tito and Rodriguez 1988, SRD 1994, Bodley pers. comm.), but others feel that this is unnecessary (e.g. Valera pers. comm.). If dewinging is required, the simplest method is to break the wing off inside the pod at the time of extraction (Bodley pers. comm.). Alternatively, wings can be broken off by rubbing and the seed cleaned to remove empty seed, parts of seed wings and other debris by using a winnowing chamber (Busby 1967).

All authors recommend the drying of seeds to prevent rot and a rapid fall in viability with storage. However, different techniques are suggested. Bauer (1987b) believes 1-2 days in the sun is usually sufficient, making sure that the seeds are 'stirred' 3-4 times per day to keep them separate. In the Philippines, seeds are dried in the sun daily for a week before storage (Valera pers. comm.). In contrast, Bodley (pers. comm.) leaves seeds to dry in a well-ventilated room for 1.5-2 months before refrigeration. Busby (1967) recommended that seeds should be dried in the sun or a grain drier to a moisture content of 8-10%. Neil (1986) found that seed dried to 8% moisture content in an oven at 35-40°C had very high viability after storage.

Tito and Rodriguez (1988) describe the treatment of seeds with Aldrin to prevent insect or fungal attack. In Fiji, a fungicide powder is applied (Oliver 1992, SRD 1994). Pre-treatment with insecticides or fungicides may not be required (Bodley pers. comm.).

Table 4.2. Number of mahogany seeds (from plantation trees) per kg

Seeds kg^{-1}	Country	Source
1257	Peru	Tito & Rodriguez (1988)
1300-3700	World-wide	Evans (1992)
1430-2070	Central America	Lamb (1966)
1500-2000	World-wide	Lamprecht (1989)
1600-2300 [a]	Philippines	ERDB (1996)
1650	India	Streets (1962)
1810 [b]	Puerto Rico	Bauer (1987b)
2000	Puerto Rico	Holdridge & Marrero (1940)
2000-2500	Indonesia	Soerianegara & Lemmens (1994)
2300 [c]	Indonesia	Bramasto & Harahap (1989)
2400	Solomon Islands	Chaplin (1993)

Note: it is unclear if seeds were dewinged before weighing, except for:
[a] Seeds with wings and [b] seeds without wings.
[c] *Swietenia macrophylla* x *S. mahagoni*

There is a large variation in reported seed weights (see Table 4.2). Most sources fail to specify whether seeds have been dewinged and/or dried before weighing. Such treatments may partly explain variation in the number of seeds per kilogram. Variation in weight may also be caused by variations in the size of capsules from which the seeds are taken. Busby (1967) found that seeds taken from capsules more than 14 cm long were nearly 3 times heavier than seeds taken from capsules less than 11.5 cm long. The age or size of the tree will probably affect the size of the capsule and thus weight of the seed.

4.3 Seed viability

4.3.1 Viability of fresh seed

The viability of 'fresh' mahogany seed is around 80-90% (see Table 4.3). Variations in seed viability could be the result of the different provenances used in the experiments, but are just as likely to be the result of differences in:

- Definition of 'fresh' seed. No authors report the length of delay between seed collection and sowing. Yet even a short delay between collection and sowing may be signifcant: de Araujo (1971) found that viability declined from 90% to 66% in just 30 days.
- Seed selection procedure. Seed viability varies with seed size. For example, Chinte (1952) showed that 'large' seeds have a 12% higher germination percentage and produce healthier, faster-growing seedlings with better developed root systems than small seeds. Busby (1967) found that germination was 90% or more for seeds from capsules more than 14 cm in length but less than 70% for seeds from capsules less than 11.5 cm. Tito and Rodriguez (1988) found that the proportion of viable seeds in some capsules fell to 68%.
- Sowing conditions. Germination tests may be carried out under laboratory or nursery conditions. In Vanuatu, seed viability under nursery conditions fell to 85% (Neil 1986). In Fiji, seed viability in nurseries fell to 60-70% on average and lower still in nurseries located on clay soils, because surface baking after heavy rains produced a cement-like crust on the beds (Busby 1967).

With no standardised experimental procedures, it is difficult to come to any conclusions regarding variation in seed viability between countries.

Table 4.3. Viability of fresh mahogany seed

Viability (%)	Country	Source
90	Malaysia	Streets (1962)
90	Philippines	ERDB (1996)
90*	Vanuatu	Neil (1986)
90*	Fiji	Busby (1967)
85-95	Brazil	de Araujo (1971)
80-90	Martinique	Marie (1949)
80-85	Philippines	Valera (pers. comm.)
78	India	Streets (1962)

* Under laboratory conditions

4.3.2 Viability of stored seed

Mahogany seed is orthodox rather than recalcitrant (Tompsett 1994). In general, mahogany seed will not retain an acceptable level of viability if stored at room temperature and humidity for more than about 3 months (see Table 4.4). Seeds of the Meliaceae are susceptible to chilling damage below about 16°C when they are moist (Tompsett 1994). If they are to be refrigerated they must therefore be dried first. Species with oily embryos such as mahogany have optimum longevity at lower moisture contents than species with starchy embryos. It appears that dried seed, refrigerated to 2-8°C and maintained at constant humidity, may maintain viability for a year or more (see Table 4.4).

Table 4.4. The effect of different storage conditions on mahogany seed viability

Temp. (°C)	Humidity (%)	Container type	Time (months)	Viability after storage	Source
-20	5	nda	24	100%	Tompsett (1992)
0	nda	nda	>3-4	Sharp decline	Vivekanandan (1978)
2-5	45	nda	<12	Acceptable	Soerianegara & Lemmens (1994)
2-5	nda	nda	12	Good	Lamprecht (1989)
4	3	Polythene bags	3	Very good	Oliver (1992), SRD (1993)
4	3	Polythene bags	3-12	Variable	Oliver (1992), SRD (1993)
4	3	Polythene bags	>12	Poor	Oliver (1992), SRD (1993)
4	8-10	Polythene bags	12	90%	Busby (1967)*
4	nda	nda	12	>80%	Bauer (1987b)
5	8	Polythene bags	24	75-80%	Neil (1986)*
8	nda	Sealed or sawdust	18	Maintained	ERDB (1996)*
15	nda	nda	>3-4	Gradual decline	Vivekanandan (1978)
15.5	50	nda	6	98%	Anon. (1956)
30	nda	nda	>3-4	Sharp decline	Vivekanandan (1978)
Room	Room	nda	1	66%	Anon. (1973)
Room	Room	nda	>1	Poor	Tito & Rodriguez (1988)
Room	Room	nda	2-3	Lost	ERDB (1996)*
Room	Room	nda	3	79%	Anon. (1956)
Room	Room	nda	3	Very good	Busby (1967)*
Room	Room	nda	<4	Slight decline	Frith (1958)
Room	Room	nda	4-6	Rapid decline	Frith (1958)
Room	Room	nda	6	40%	Busby (1967)*
Room	Room	nda	>6	0%	Frith (1958)
Room	Room	nda	8	0%	Anon. (1956)
Room	Room	nda	9	10%	Busby (1967)*
Room	Room	nda	12	>80%	Bauer (1987b)
nda	4-5	Sealed containers	8	Good	Lamb (1966)

* These authors also report viability of fresh seed (see Table 4.3) nda = no details available

A number of authors believe that seed sealed in containers (polythene bags, jars or demi-johns) retains its viability for longer owing to maintenance of constant humidity (Lamb 1966, Chable 1967, Boontawee 1996). Transparent containers enable the contents to be checked easily. If seed starts to turn a dark brown colour, deterioration is taking place and seeds must be re-dried and re-sealed (Lamb 1966). However, an experiment carried out by Holdridge and Marrero (1940) indicates that the type of container used makes little difference (see Fig. 4.1). Storage temperature was found to be a more important variable than type of container.

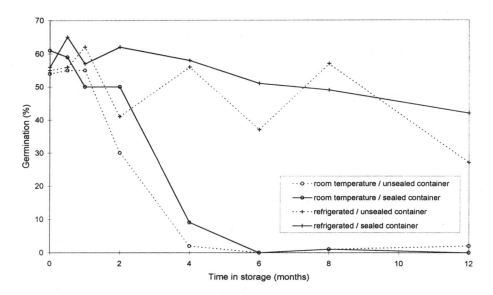

Fig. 4.1. Mahogany seed viability after storage (after Holdridge and Marrero 1940).

A precise interpretation of reported results is precluded by the omission of certain key data. Few authors report all relevant details (i.e. temperature, humidity *and* storage container). The definitions of 'room' temperature and humidity will vary according to local climate, the design of building and the type of container in which seed is stored. Most authors do not provide details of seed viability before storage. It appears that many have assumed unstored seeds have 100% viability (probably a false assumption; see Section 4.3.1). Poor viability post-storage may therefore be a reflection of poor viability pre-storage, making it impossible to assess the decline in viability at the time of the first measurement.

According to conventional wisdom, storage temperatures should not fall below freezing (Lamb 1966). However, this is not the case for seeds that have been dried sufficiently. Tompsett (1994) found that *S. macrophylla* seeds dried to a moisture content of 5% and stored at a temperature of minus 20°C maintained a viability of up to 100% over a 2 year period. It was predicted that *S. humilis* seed dried to 2% and stored at an optimal temperature of minus 13°C could survive for 266 years.

Interim seed storage may be important if seed sources are distant from proper storage facilities. Seeds of Meliaceae can be stored in the short term at a temperature of 18°C to avoid chilling damage, with ventilation (Tompsett 1994). Seed should be transported in boxes with small holes, packed into containers with a high air to seed ratio. This minimises dangers of both anoxia and desiccation (due to over-ventilation).

Pre-treatment with insecticide may reduce damage to unrefrigerated seed caused by weevils and other pests; however, the extent of insect predation problems is generally not reported. Busby (1967) found that pre-treatment with a mercurial seed-dressing maintained germination capacity at 90% for 12 months. The use of this dressing was described as a 'major breakthrough' at the time, but given recent concerns over use of heavy metals, such treatment would now be unacceptable in many countries.

It appears that no treatment of seed is required to break dormancy after storage (Chaplin 1993). Trials showed that germination was significantly improved by a 24 hour soaking under laboratory conditions, but untreated seeds germinated significantly better under nursery conditions (SRD 1993).

4.4 Summary

- Seed production under plantation conditions usually begins by the age of 15 years. Production fluctuates considerably from year to year. Hurricane damage to mahogany crowns may delay seed production for several years, causing periodic seed shortages.
- To obtain a high proportion of germinating seed, capsules should be harvested from the tree towards the middle or end of the fruiting season. The viability of fresh seed may be as high as 95%, depending on seed selection criteria.
- If seed is to be stored, seed viability can be enhanced by drying, sealing in containers and refrigerating (to 2-8°C). If seed is dried to a moisture content of 5% or less, deep freezing (at minus 20°C) will maintain a high viability for at least 2 years and possibly many decades.
- Chemical treatment of seeds before storage may help to maintain long-term viability, but treatment to break dormancy does not appear to be necessary.

Chapter Five
Nursery Techniques

5.1 Sowing

Mahogany can be sown in beds to produce bare-rooted planting stock because the species is relatively hardy (Evans 1992). Raising stock in containers may give a more fibrous rooting system and better results after planting out, but at an extra cost. The decision between beds and pots depends principally on local practices and available resources. Planting site conditions may also be taken into account. For example, it may be preferable to use bare-rooted mahogany stock in wet areas and container stock in dry areas (Bodley pers. comm.).

5.1.1 Sowing in beds

No special preparation of nursery beds is required for mahogany. Light, well-drained soils give the best results (Lamb 1966) but the species can grow satisfactorily in sub-optimal conditions. Mahogany seeds are large and easy to sow in uniformly spaced drills, which facilitates weeding (Holdridge and Marrero 1940).

Recommended mahogany sowing densities vary widely between countries and are a reflection of available nursery space, nursery management costs, quantities of available seed, seed viability, seedling growth rates, desired size of planting stock and standard nursery practices. The examples below therefore define the range of suitable sowing densities rather than an optimal value.

In the Philippines, seeds are sown at 5 x 10 cm and seedlings transplanted to plastic pots when they are 15 cm high (Anon. 1955a, ERDB 1996). Streets (1962) suggests a spacing of 10 x 20 cm prior to transplantation. Chable (1967) reports transplanting from 10 x 10 cm to 45 x 45 cm for more rapid development. Transplantation is carried out after 3 months (Flinta 1960).

To obtain seedlings of 30-60 cm in height, a square spacing of 10-15 cm is commonly recommended (Holdridge and Marrero 1940, Marie 1949, Anon. 1959, Neil 1986). Streets (1962) and Lamb (1966) suggest a somewhat wider spacing of 15 x 30 cm. In Fiji (Nukurua nursery), seedlings are spaced at 10 x 30 cm (Evo pers. comm.). To obtain seedlings of 100 cm in height, a spacing of 20 x 30 cm is suggested (Bauer 1987b).

If seedlings are selectively removed (for planting out), seeds can be sown significantly closer. In St. Lucia seeds are sown 2.5-5 x 15 cm apart (Bodley pers. comm.). As the seedlings grow, the biggest are dispatched, liberating the smaller seedlings, which then grow to a better size. Staggered production of planting stock is useful where planting seasons are long and field staff in short supply.

5.1.2 Sowing in containers

Mahogany container stock is used in regions where bare-rooted planting stock has a low survival. In Western Samoa and some nurseries in the Philippines, mahogany seeds are sown directly into pots, with one seed per pot (Oliver 1992, Valera pers. comm.). It may be necessary to sow more than one seed per pot if the germination percentage is low (Flinta 1960, Lamb 1966). If necessary, pricking out should take place at an early stage (see Section 5.1.3).

Not all authors feel that the advantages of a more fibrous rooting system are sufficient to merit the use of containers. Work in Fiji indicates that strong development of the tap root should be encouraged if plantations are to be wind firm (Singh pers. comm.). Some believe that containers should never be used because they restrict the strong (and healthy) development of the mahogany tap root (de Araujo 1971, Anon. 1973). Other problems, such as spiralling roots in wide pots, have been noted (Oliver 1992). Mahogany seedlings at Carua, Brazil, were established easily with bare-rooted plants and containers were found to be unnecessary (Anon. 1961).

5.1.3 Sowing in trays

Sowing seeds at high densities in trays of sieved top-soil has given good results in the Solomon Islands (Chaplin 1993). The use of sawdust to cover the seeds should be avoided since it makes it difficult to maintain correct moisture levels and can cause the spread of fungal infection. Seedlings are ready for pricking out when the first seed leaves appear. Immediately after pricking out, 70% shade is required, which is reduced after about a week to 30% and finally removed 2-3 weeks later (Anon. 1985).

Lamb (1966) recommends pricking out 4-5 days after germination, while the mahogany plants have only 2 small leaves. However, others suggest a longer period between germination and pricking out: 6-10 days (Anon. 1985) or even 21 days (Chai and Kendawang 1984). As a general rule, the seed should still be attached to the shoot and supplying nutrients at the time of pricking out (Flinta 1960, Oliver 1992).

5.1.4 Optimal sowing depth and seed orientation

Research into the optimal sowing depth of mahogany seed has produced conflicting results. Mondala (1977) sowed seeds at a range of depths from 2 to 7 cm and found no statistical difference between the treatments. Schmidt (1974) also showed that sowing depth had no effect on germination, but that it did affect the height of the seedlings. Seedlings sown at a depth of 1 cm grew significantly taller, and had a greater dry weight than seedlings sown at 8 cm. Busby (1967) argues that sowing at greater depths reduces germination considerably. In contrast, Chinte (1952) reports that a depth of 4-8 cm is favourable for seed germination, with best results at 8 cm.

A survey of the literature revealed that most authors in fact recommended a sowing depth of less than 2 cm (Noltee 1926, Holdridge and Marrero 1940, Flinta 1960, Lamb 1966, Anon. 1973, Schmidt 1974, Oliver 1992, Chaplin 1993, Evo pers. comm., Valera pers. comm.). Some favoured a depth of 2-4 cm (Anon. 1955a, 1959, Busby 1967, Bauer

1987b, Bodley pers. comm., Lamprecht 1989, Chaplin 1993). Only a few suggested sowing at greater depths (Marie 1949, Chinte 1952). Given that mahogany seed is attached to a large wing and is relatively short-lived, it is unlikely to be buried under natural conditions. The seed may therefore be poorly adapted to germinate at depth. However, in an attempt to simulate natural regeneration, Tito and Rodriguez (1988) found that sown seeds had a significantly higher germination percentage than seeds scattered on the soil surface.

Evidence that orientation of the seed at the time of sowing may affect germination is contradictory. By planting the seed with its long axis vertical, Neil (1986) and Valera (pers. comm.) claim that deformities of the developing radicle are reduced. But Tito and Rodriguez (1988) found that seeds planted on their side had a significantly higher germination percentage than seeds planted vertically. Others also suggest that seeds should be planted horizontally, either flat or on one edge with the germinating side pointing up (Noltee 1926, Lamb 1966). In their germination experiments Parraguirre and Camacho (1992) sowed mahogany seeds with the wing-end pointing towards the surface at an angle of 45°. This method is thought to facilitate the emergence of the seedlings and is common practice in Quintana Roo, Mexico.

Mondala (1977), who planted seeds in 3 sowing positions (wing pointing up, flat or down), found no statistical difference between the treatments. Research during the 1960s in Fiji also showed that the orientation of the seed has no effect on germination (Busby 1967). In a more recent Fijian experiment, 4 different orientations of seed were tried: each one had over 91% germination, with no significant difference between them (SRD 1995).

5.1.5 Germination

There is considerable variability in timing and duration of germination of viable mahogany seeds (see Table 5.1). A number of possible causes of this variability have been suggested:
- Seed freshness. Work carried out by Bauer (1987a) indicates that use of old mahogany seed will extend the period of germination and that this factor alone may explain most of the variability.
- Soil type. Mahogany seeds have a high fatty content and require more oxygen during germination than starchy seeds (Crocker 1948). Germination may therefore be faster in well-aerated soils.
- Soil moisture. Seeds require an abundant supply of moisture (Anon. 1955a, Lamb 1966). Germination may therefore be faster in well-watered soils, although soils should not become waterlogged due to risk of de-oxygenating the soil and 'damping off' (Anon. 1955a).
- Soil compaction. The formation of a crust on the soil above the seeds will cause the shoot to coil under the surface and emergence will be delayed (SRD 1995).

Table 5.1. Timing and duration of germination of mahogany seed

Start (days*)	End (days*)	Progress during the germinating period	Notes	Source
5	40	50% of seeds germinated by day 22 and 75% by day 25. Peak rate recorded on day 28	Seed fresh (15 days old)	Parraguirre & Camacho (1992)
10	30	nda	Figures based on wide experience	Lamb (1966)
13-17	23-27	nda	-	de Araujo (1971)
15	49	nda	Seed not fresh	Bauer (1987b)
nda	28	nda	Average of 1319 sowings	Marrero 1949
nda	nda	Most seeds germinated during days 8-14	Seed fresh and mature	ERDB (1996)
nda	nda	Peak germination rate observed during days 14-16	-	SRD (1993)
nda	nda	Most seeds germinated by day 21	-	Busby (1967)

* After sowing   nda = no details available

5.2 Vegetative propagation

5.2.1 Propagation by leafy cuttings

Various species of the Meliaceae have been successfully propagated by rooting leafy cuttings (Leakey *et al.* 1982, Newton *et al.* 1993b, 1994). Successful results have been obtained with different propagation systems, including traditional mist propagators (Howard *et al.* 1988) and low-technology non-mist propagators (Leakey *et al.* 1990, Newton *et al.* 1994). However, if *Swietenia macrophylla* is to be propagated on a commercial scale, detailed information is required on the appropriate treatments which should be applied to both the stockplants and the cuttings, in order to obtain consistently high rooting success. Relatively few experiments have been undertaken with mahogany.

In a preliminary investigation using non-mist propagators, percentage rooting of mahogany cuttings was found to be highest when a rooting medium with a high proportion of sand was used. A rooting proportion of over 60% was achieved with 75:25 sand:gravel (Newton *et al.* 1993b). The concentration of IBA applied to the base of the cuttings was found to have only a slight effect on rooting. In these experiments, the cuttings were relatively slow to root (11 weeks). SRD (1994) found that 80% of shoot cuttings taken from 18-month-old seedlings rooted if IBA is applied, with 60% recorded when no IBA was applied. Ompad (1947) showed that mahogany cuttings between 6.5 and 7.5 cm seemed to showed the highest survival rates, although differences between all other size classes was negligible.

SRD (1994) also investigated stooling techniques. Seedlings were cut at 4 cm, 8 cm, 12 cm above ground. Seedlings cut at 12 cm produced the most shoots over a 22 week period. The rooting percentage of cuttings taken from all stools was around 80%. Chinte (1952) briefly studied the effects of coppicing. From a sample of 295 tree stumps, 81% produced shoots; the younger trees had a greater sprouting ability than older ones. The optimal stump size for coppice production has yet to be determined.

The rooting ability of cuttings taken from coppice regrowth is generally much higher than of cuttings taken from mature branches (Leakey *et al*. 1994). Further research is therefore required to define appropriate management of stockplants for production of coppice shoots suitable for propagation by cuttings. Possible approaches include pollarding branches or slashing the bark to induce coppice shoot formation without felling the tree.

5.2.2 *In vitro* micropropagation

In vitro techniques have the advantage of very high multiplication rates, but are generally costly in terms of training, expertise and equipment. In addition, adaptation of planting stock to field conditions may be difficult. Micropropagation of *Swietenia* spp. has already met with some success and has provided a means of propagating plant material extremely rapidly (Evans 1992). Details of micropropagation techniques applied to *S. macrophylla* and other related Meliaceae are given by Lee and Rao (1988), Maruyama *et al*. (1989) and Newton *et al*. (1994). One of the potential applications of micropropagation techniques is the multiplication of pest-resistant genotypes, which offers a possible approach to overcoming problems of shoot borer attack in mahogany (Newton *et al*. 1994).

5.3 Maintenance

Mahogany seedlings will survive without much assistance (Busby 1967, Francis pers. comm.), but good maintenance is essential for the production of healthy and fast-growing plants. Watering, weeding and fertilising regimes are designed primarily in response to local conditions so it is unhelpful to cite general guidelines. In most nurseries, mahogany seedlings are treated like any other species and standard practices are followed. The situations for which standard practices may not be suitable are described below.

5.3.1 Shading

Although mahogany is a light-demanding species, most nursery managers provide shade for young seedlings (see Table 5.2). It is not clear from these practices whether mahogany seedlings are actually intolerant of exposed conditions in the nursery, or whether shade is used simply to reduce watering requirements. The Fijian Forestry Department is unusual in leaving nursery beds unshaded (Oliver 1992, Evo pers. comm.). Seedlings are planted at the coolest time of year when the risk of desiccation is lowest. From his observations in Fiji, Evans (1992) concludes that mahogany does not need shading and can be grown from the time of germination in an open bed.

Table 5.2. Reported shading regimes for mahogany seedlings

Country (region)	Depth of shade	Period of shade	Source
For seedlings immediately after germination			
Fiji *	50% shade (bamboo slats)	6 weeks	Busby (1967)
Indonesia (Java)	Light shade	12-16 weeks	Soerianegara & Lemmens (1994)
Malaysia (Penins.)	100% reduced to 50% shade	heavy for 8-12 weeks; light for 4 weeks	Streets (1962)
Peru	nda	8-12 weeks	Flinta (1960)
Philippines (Luzon)	(translucent plastic)	8-12 weeks	Valera (pers. comm.)
Solomon Islands	70% reduced to 30% shade	heavy for 2 weeks; light for 6-8 weeks	Oliver (1992)
Western Samoa	30% shade	whole period in nursery bar last 2 weeks	Oliver (1992)
World-wide	Light shade	12-24 weeks	Lamprecht (1989)
For seedlings immediately after transplantation			
Malaysia (Penins.)	100% reduced to 50% shade	heavy for 2-3 weeks; light for 2-3 weeks	Streets (1962)
St. Lucia	(fine mesh netting)	2 weeks	Bodley (pers. comm.)

* Mainly in sandy lowlands where soil temperatures were high   nda = no details available

5.3.2 Root pruning

Root pruning is used to reduce tap root development and produce a more fibrous rooting system. The advantages of root pruning are disputed by Nobles and Briscoe (1966), who found that mahogany seedlings did not benefit. However, most sources record positive results. Djapilus (1988) notes that bare-root mahogany stock after pruning was superior to container stock. Busby (1967) found that root pruning improved survival rates in the field: even with adverse conditions after planting, 80% of root-pruned stock survived, whereas on average only 50-60% of unpruned stock survived. Data produced by Marin (1963) suggest that root-pruned planting stock has a lower mortality 6 months after planting, although the claim that root pruning also improved height growth was not supported by his data.

A detailed experiment on the effects of root pruning on mahogany seedlings was carried out by Asiddao and Jacalne (1959). Roots were pruned at about 10 cm below the surface of the nursery bed. Pruning was found to slow the height growth of the seedlings. For seedlings root pruned at 30 days, heights were 84% of the control (unpruned) seedlings after 16 weeks; for seedlings root pruned at 60 days, heights were 74% of the control after 16 weeks. It was recommended that pruning be delayed until the flush of young seedlings is almost over. With care, root pruning causes little mortality: in this experiment only 5 plants out of 300 died. A more fibrous root system was observed to develop and a significant difference was found between the weights of control and root-pruned seedling root systems over a 3 month period, regardless of timing of root pruning.

Because of the tendency of mahogany seedlings to develop a long tap root, root pruning is practised in many mahogany-growing countries. In St. Lucia the tap root is pruned to 15 cm after lifting (Bodley pers. comm.). Root pruning is done just 2 weeks before lifting in the Solomon Islands (Anon. 1985, Chaplin 1993). In Fiji, Busby (1967) recommended that mahogany seedlings are root pruned 4-6 weeks before lifting, when the seedlings are around 45 cm high. If root pruning is carried out any earlier it may have to be repeated, but if left any later there is a risk that the seedling will not recover in time for planting out. However, practices in Fiji have changed and root pruning is now carried out 3 months before lifting to allow seedlings a chance to recover (Evo pers. comm.). Root pruning in Fiji may be phased out altogether because of the apparently greater susceptibility of pruned seedlings to windthrow (see Section 12.2.2).

The use of container planting stock removes the necessity for root pruning. Suitable containers for growing mahogany have holes in the bottom to allow roots to grow through, thus minimising root coil (SRD 1995, Bodley pers. comm.). As long as containers away from the ground, roots are air pruned and no extra work is required.

5.3.3 Controlling pests and diseases

A number of pests and diseases have been associated with mahogany in nurseries around the world (see Tables 5.3a and 5.3b). Of those reported, only *Hypsipyla* spp. appears to be specific to the Meliaceae. Most pests and diseases can be prevented or contained by maintaining reasonable standards of hygiene and a reactive watering regime, combined with careful application of pesticides and fungicides where necessary. Evidence of severe damage to mahogany seedlings is rare and suggested methods of control have been used effectively when problems arise.

5.4 Seedling growth and recommended size for planting out

The growth rates of mahogany seedlings vary considerably, depending on soil fertility and nursery maintenance. Seedlings average 30 cm after 4 months and 60 cm after 6 months in Puerto Rico (Holdridge and Marrero 1940). Similar growth rates were observed in Solomon Island nurseries, with seedlings reaching 30-50 cm after 4 months (Oliver 1992). Lamb (1966) and Rodriguez (1996) are more optimistic, predicting a height of 60-90 cm in 6 months. Height growth of over 1 m in 6 months has been observed (Marrero 1942).

In general, seedlings may be planted out with little mortality once the stem base has become woody (Streets 1962). Busby (1967) suggests a root collar diameter of 0.8-1.3 cm. Lamb (1966) believes that seedlings need a root collar diameter of 1-2 cm to survive the disruption of transplanting. The leader should be strong, not tender (Marie 1949), and fresh shoots should have had a chance to mature and harden (Evo pers. comm.).

Results from research trials which have attempted to define the optimal height for bare-rooted planting stock are inconclusive. Marrero (1942) found that 60-100 cm plants had a significantly higher survival rate than larger (100-160 cm) plants. Chinte (1952) found that plants 20-35 cm tall had a 60% survival rate, whereas plants 70-85 cm tall had a 95% survival rate. However, in this experiment there was a 3 week delay between lifting and transplanting and the figures may reflect the ability of larger plants to survive long periods in storage. Other researchers recommend heights of anything from 60 cm to 150 cm (e.g. Anon. 1955a, Perera 1955, de Araujo 1971, Bauer 1987b). In contrast, Holdridge and Marrero (1940) found little difference between the survival and growth of 30-50 cm tall plants and 120-150 cm tall plants. Variable climatic conditions and nursery practices (unrecorded) may explain much of the difference between experiments.

Table 5.3a. Diseases affecting mahogany seedlings in the nursery

Species	Nature of attack and/or symptoms	Suggested control	Country	Source
Botryodiplodia theobromae (Stem rot)	Affects the region just above the root collar on seedlings up to 2 months old, leading to wilting, browning and curling of leaves and death of the seedling. Afterwards black, ovate fruiting bodies appear on the stem	Dust the diseased bed with pentachloro-nitrobenzene (PCNB) powder at 10-20 g m^{-2} and mix in with the soil. Avoid use of soil with high water-holding capacity, overwatering and dense sowing	Philippines	Dayan (no date), de Guzman & Eusebio (1975), Roldan (1941)
Corticium koleroga (Thread blight)	nda	Use Captan or a copper fungicide	Fiji	Busby (1967)
Fusarium spp. (Damping-off fungus)	Causes leaves to turn reddish or purplish and suppresses growth	Disinfect the soil before seeding with a dilute emulsion of Captan (3-6 litres m^{-2} of soil). The mahogany seeds should be treated by drenching with Thiram or Delsene MX (fungicides)	Philippines	Dayan (no date)
Pellicularia spp. (Thread blight)	Recorded during hot and humid spells	Use Captan or a copper fungicide	Fiji	Busby (1967), SRD (1994)
Rhizoctonia solani (Damping-off fungus)	Causes leaves to turn reddish or purplish and suppresses growth	Disinfect the soil before seeding with a dilute emulsion of Captan (3-6 litres m^{-2} of soil). The mahogany seeds should be treated by drenching with Thiram or Delsene MX (fungicides)	Philippines	Dayan (no date)
Sclerotium spp.	Mainly affects seedlings up to 2 months old. White, mycelial felts may be observed in affected parts of the nursery bearing brown, oblong, sclerotial fruiting bodies. Causes collar rot	Dust the diseased bed with pentachloro-nitrobenzene (PCNB) powder at 10-20 g m^{-2} and mix in with the soil. Avoid use of soil with high water-holding capacity, overwatering and dense sowing	Philippines, India	de Guzman & Eusebio (1975), Sankaran *et al.* (1984)
Species unknown (Damping-off fungus)	Rotting of tissue at the root collar causes seedlings to topple over	Use Benlate. Avoid overwatering, overshading and underventilating	Western Samoa	Evans (1992), Oliver (1992)

Table 5.3b. Pests affecting mahogany seedlings in the nursery

Species	Nature of attack and/or symptoms	Suggested control	Country	Source
Acrocerops auricilla (Leaf miner)	White, oval blotches about 11x7 mm in size appear on leaf. Mined tissue turns red and withers	nda	Philippines	Dayan (no date)
Diaprepes abbreviatus (Sugar cane weevil)	Partial defoliation is caused by adult. More serious damage is caused by the larva, which feeds on the root stalk just below the ground surface, sometimes causing mortality	Use commercial insecticides	Puerto Rico	Bauer (1987b), Evans (1992)
Helopittis antonii (Leaf bug)	Feeds on sap, killing leaf tissues and sometimes young shoots	nda	Philippines	Dayan (no date)
Hypsipyla spp. (Shoot borer)	Larva damages or destroys apical shoot	Apply Orthene (systemic insecticide) weekly. Orthene should also be applied to seedlings a few days before planting out to prevent transfer of borer populations into the field	Solomon Islands	Wilson (1986), Chaplin (1993)
Xyleborus abruptoides (ambrosia beetle)	Found in the stems of dying nursery seedlings	Avoid water or nutrient stress	Fiji	Gibson (1975)
Xyleborus caffeae (ambrosia beetle)	Bores into the main root or root collar	nda	Indonesia (Java)	van Hall (1919), Kalshoven (1926),
Slugs and snails	Cause branching of seedlings due to removal of leaves	Use poisoned slug bait	St. Lucia, Indonesia	Bramasto & Harahap (1989), Bodley (pers. comm.)

nda = no details available

There is also considerable variation in the heights of planting stock preferred by forest managers (see Table 5.4). The main advantage of using large planting stock is the reduced cost of vine cutting and weeding on planting sites (Holdridge and Marrero 1940). However, large planting stock is also prone to root damage and water stress after planting and many (e.g. Marrero 1942, Anon. 1961, Lamb 1966) recommend cutting back large stock for this reason (see Section 5.5.2). Forest managers must also take into account the additional nursery, transport and planting costs associated with growing larger seedlings. Beating up after 6-12 months may prove to be a cheaper proposition than producing and handling plants which have the highest chance of survival.

Table 5.4. Heights of bare-rooted planting stock preferred by forest managers

Height (cm)	Country	Source
30-45	Fiji	Evo (pers. comm.).
30-60	India	Streets (1962)
40	St. Lucia	Bodley (pers. comm.)
60	Fiji	Busby (1967)
75	Solomon Islands	Anon. (1985)
100	Central America	Lamb (1966)
100	Honduras	Chable (1967)
100-120	Martinique	Marie (1949)

The shoot borer can be chemically controlled in the nursery but control in the field is expensive and not always successful (see Section 11.2). Weaver and Bauer (1986) therefore recommend the use of large planting stock, which will have less chance of being attacked by the shoot borer before reaching its desired unbranched height. However, there is no evidence that large planting stock is less prone to attack. In fact, Bauer (1987a) indicates that shoot borer attack was highest on plants which were over 1.5 m when planted out. Vigorous growth after planting, which may promote recovery after shoot borer attack, is probably a more important consideration than the height of planting stock (see Section 11.1.1.3).

The use of container planting stock limits the choice of sizes for planting out, since the container prohibits proper development of roots. Container stock more than 100 cm high is likely to have a coiled root system and it may be preferable to take stumps (see Section 5.5.2). Planting out of container stock in the Philippines is done when seedlings are 50 cm high (ERDB 1996).

5.5 Preparation of bare-rooted stock for planting out

Nursery beds should be thoroughly watered before seedlings are lifted to avoid straining or breaking the small, fibrous roots (Anon. 1955a, Busby 1967). Lifting and subsequent preparation of bare-rooted stock must take place as close to the planting date as possible to minimise the risk of desiccation.

5.5.1 Leaf stripping

Mahogany seedlings are commonly stripped of some or all of their leaves to produce striplings. The technique is used for many species of the Meliaceae (Evans 1992). Striplings have lower transpiration rates and are therefore less susceptible to drought effects after planting out. Seedlings which have not been stripped of leaves have been used in Sri Lanka and French colonies (Anon. 1959). However, they must be planted during times of slow growth and not while flushing, or the terminal bud will die.

Lamb (1966) recommends the removal of all leaves, although he stresses that the stem should not be cut back, nor the terminal bud damaged. This technique is used in St. Lucia (Bodley pers. comm.). In nurseries in Honduras, a 'haft' of leaves was left at the top of the stem to protect the young bud (Chable 1967). The practice in Fiji is to strip off all but the top pair of leaves on the leading shoot (Oliver 1992, Evo pers. comm.). In the Solomon Islands, the leaf area of the seedling is reduced to the top one or two pairs, 14 days before out-planting (Oliver 1992).

Striplings are preferred by Busby (1967) because of higher survival rates in adverse climatic conditions, although it is noted that, under normal conditions, height increment is greater for plants with all their leaves.

5.5.2 Cutting back

When the nursery stock is oversized, planting stock should be cut back to 45 cm a month before planting date (Lamb 1966). Cutting back should be carried out in advance of lifting to avoid stressing the plant. After planting, the leader is re-formed quickly and the stump is absorbed. Marrero (1942) found that the survival of oversized (100-160 cm) stock was significantly improved by cutting back. The more severe the cut, the better the survival: stock cut back to 10 cm had an equivalent survival to uncut regular stock (60-100 cm). However, for regular stock cutting back to 10 cm gave significantly lower survival rates.

In some situations, bare-rooted plants are cut above and below the root collar to produce stumps. The main advantage of using stumps is that they are less bulky and therefore easier and cheaper to transport. Stumps are planted in crowbar holes (Jacalne *et al.* 1959) which makes planting quick and easy. Stumps may be able to survive a longer delay between lifting and planting.

Research on stump sprouting was carried out by Jacalne *et al.* (1959). Tap roots were cut back to 25 cm (side roots removed) and the stem cut 2-3 cm above the root collar. The stem diameter at the root collar was 1-1.5 cm. After planting, sprouting occurred over a 2-41 day period, although for smaller diameter stumps the period appeared to be shorter. Planting in the middle of the rainy season appeared to give the best results, but even then only around 50% of stumps sprouted. Smaller diameter stumps appeared to give slightly better growth and survival. Seedlings gave better survival and growth over a 2 year period to all stumps. In a different experiment, Perera (1955) found that stumps did not survive, but his results may have been different if his plots had been properly weeded.

If stumps are to be taken, growing stock should be about 150-200 cm high and with a root collar diameter 'of at least a thumb's width' (Lamprecht 1989). Recommended stump dimensions are: stem 20 cm and root 20-40 cm, although optimal size has yet to be determined. In Curua, Brazil, stumps were taken from plants over 1 m tall (Anon. 1961).

5.5.3 Storage

Experience in Fiji indicates that bare-rooted mahogany plants are hardy enough to survive for a few days in the open when stored under shade (Oliver 1992). However, without protection from desiccation, fibrous roots are destroyed. Survival at the planting site can be enhanced if seedlings are treated more carefully and roots kept moist. Reported methods include: wrapping the roots of the bundles in dry grass and soaking (Bodley pers. comm.); putting bundles inside cloth bags with wet sphagnum moss (Bauer 1987b); putting bundles in wet sacks (Evo pers. comm.); putting individual plants in polythene bags, which are in turn placed in wet sacks to keep them cool (Anon. 1960, Lamb 1966). Bundles, bags or sacks should then be kept in the shade until planting out (Lamb 1966).

It is recommended that plants should be stored for a maximum of 2 days (Bodley pers. comm., Evo pers. comm.). However, Anon. (1960) and Lamb (1966) found that plants can be kept for up to 7-10 days without deteriorating, using their own storage techniques (see above). Perera (1955) noted that survival rate for large (135-150 cm) mahogany striplings after 2 days' transport and 16 days' storage in shaded trenches was over 90%. Chinte (1952) documents a high survival rate, even after 3 weeks in storage, for plants where roots are properly protected.

5.6 Summary

- Mahogany seeds are large and easily sown in nursery beds or pots. Planting in nursery beds tends to encourage tap root development, whereas containers encourage a more fibrous rooting system. Container planting stock may be more suitable for dry planting sites and bare-rooted stock for wet sites or sites which are subjected to very strong winds.
- The sowing density in nursery beds depends on space available, the desired size of seedling for planting out and whether or not transplanting is intended. A spacing of anything between 5 and 30 cm between seeds has given satisfactory results. A sowing depth of about 2 cm is suitable.
- Mahogany may be propagated vegetatively. Stumps of seedlings and young trees are able to coppice, providing a source of new shoots suitable for propagation by leafy cuttings. *In vitro* techniques have been successfully demonstrated under experimental conditions.
- Serious damage to mahogany plants by pests and diseases has rarely been recorded in the nursery. Attack by *Hypsipyla* spp. can be successfully controlled with insecticides under nursery conditions.
- Nursery plants should be kept well watered and shaded for around 12 weeks from the time of germination. Root pruning in the nursery helps to create a more fibrous rooting system. Root pruning should be carried out about 4 weeks before lifting, but should not be done while seedlings are flushing.
- Most researchers favour seedlings taller than 60 cm for planting out, but there is no agreement over the optimal size. Forest managers have found that 30-100 cm tall plants give satisfactory results. Small plants are more susceptible to weed competition but are cheaper to produce and easier to handle.

- For bare-root planting, mahogany seedlings should be stripped of all but the top pair of leaves to produce striplings. The terminal bud must not be damaged. Cutting back of oversized stock (over 100 cm) to 10-20 cm stumps is recommended and roots may also be cut back to a 20-40 cm tap root.
- Careful storage of bare-rooted planting stock is important if there is to be a delay between lifting and planting. Striplings or stumps should be wrapped, kept moist and stored in the shade.

Chapter Six
Site Selection

Descriptions of the site preferences of mahogany are largely anecdotal and the views of different authors vary considerably. Holdridge and Marrero (1940) found that 'site constitutes one of the most important factors in the success or failure of big-leaf mahogany plantations, as this species shows a marked preference for certain environmental conditions'. In contrast, Lyhr (1992) notes that 'mahogany demonstrates extreme plasticity in site preference'.

The failure of many authors to appreciate the interdependency of site variables (temperature, rainfall, soil type, soil drainage, topography, etc.) is largely responsible for the confusion encountered in the literature. Variation in the productivity of mahogany plantations is usually explained by a single site variable, or a number of variables which are considered independently. This approach may be useful in highlighting some of the important limiting factors to growth, but it precludes a full understanding of all the interactions that are occurring at a site. There have been few efforts to assess the suitability of a site for mahogany using a combination of site variables. One of the few examples comes from Guadeloupe, where Soubieux (1983) used aspect (protection from prevailing easterly wind), topography (position on the slope) and altitude (which was positively correlated with rainfall) to predict site productivity. The relationship derived was found to be significant, although given the particular climatic conditions encountered on the island, is unlikely to be useful for other locations.

Other (non-site) variables that could also explain variations in the performance of mahogany are often ignored. For example, it is possible that the different provenances used in different plantations are reacting differently to similar site conditions. The lack of a comprehensive series of provenance tests, examining interactions between site type and genetic origin, greatly limits the conclusions which can be drawn about the site preferences of mahogany. Alternatively, mahogany stands on particular site types may receive different silvicultural treatments to others. For example, stands on steep slopes may be thinned less than those on flatter ground in order to reduce soil erosion. Variations in the treatment of stands will cause variations in stand productivity.

6.1 Climate and altitude

The natural ecological range of mahogany (see Section 2.3) serves as a useful general guide for forest managers.

Data from plantation sites indicate a preference for a tropical rather than sub-tropical climate (see Table 6.1). Flinta (1960) reported that tolerable extremes of temperature for mahogany cultivation range from 12 to 37°C. The temperature range of mahogany will, to a large extent, determine its altitudinal range, reported to be 50-1400 m (Pandey 1983, Webb et al. 1984). The planting of mahogany at high (though unspecified) altitudes has been recommended in Indonesia (Thorenaar 1941) and reported in Fiji (Cottle no date), although, in the latter case, wind damage has been severe.

The minimum acceptable annual rainfall for mahogany is reported to be around 800 mm yr^{-1} (Francis pers. comm.), although this figure will vary from site to site according to the water retentivity of the soil. Early planting experiences indicated that a rainfall of about 2000 mm yr^{-1} gives satisfactory results (Holdridge and Marrero 1940). Flinta (1960) and Pandey (1983) assert that mahogany reaches its optimal development in a climate of 1000-2000 mm yr^{-1}, but may be cultivated in regions with rainfall of up to 4000-5000 mm yr^{-1}. In fact, the most high-yielding plantations are found in regions with a rainfall considerably higher than 2000 mm yr^{-1} (e.g. parts of Martinique and the Philippines).

Annual rainfall distribution is also important. A region with a high total rainfall but a long dry season may be less suitable for mahogany than a region with low rainfall and short dry season. Webb et al. (1984) suggest that mahogany can tolerate a dry season of up to 4 months. The longest reported dry season in a mahogany-growing country is 4-5 months (see Table 6.1). However, it is difficult to determine the maximum tolerable length of dry season because of the vague definition of the term. Growth rates during long dry periods are low. Mean diameter increment over a particularly dry year (1995-96) in Sri Lanka was, on average, only 2-3 mm (Thayaparan pers. comm.); 8-10 mm would be expected in a normal year. Plantations in regions with long dry seasons will not, therefore, have a high productivity.

6.2 Soil type

Soil conditions under natural mahogany stands may not necessarily reflect the most suitable soil conditions for plantation-grown mahogany, since evidence indicates that natural distribution is partly dependent on disturbance rather than soil type alone. In plantations, where most competing vegetation has been removed, optimal soil conditions may differ considerably from those within natural forest. Mahogany plantations have been established on a wide range of soil types (see Table 6.1).

Research on the effect of soils on the growth of mahogany is limited. Marrero (1942) measured the growth of 1-year-old mahogany seedlings over 4 sites with clay, silty clay loam and stony soils. It was found that significant differences in growth rates between sites could not be explained by soil type. Instead, soil protection and past cultivation appeared to be the most important variables. In the Luquillo and Rio Abajo mahogany plantations of Puerto Rico, Ewel (1963) measured tree height and associated site variables (soil pH at 35 cm, A1 horizon texture, A1 horizon depth and B horizon texture, slope, aspect and drainage) in 38 plots to test for height-site relationships. The plots each contained at least 6 mahogany trees of approximately the same age (25 years). Only one of the site variables (A1 horizon depth up to 15 cm) was found to produce a significant (at 95%) regression with tree height. An increase in soil depth of 1 cm corresponded to 25 cm increase in height growth.

Table 6.1. Climate and soil types for plantation sites in countries where mahogany has been successfully established

Country	Region	Temperature (°C)	Rainfall (mm yr^{-1})	Dry season	Altitude (m)	Soil type	Source
Belize	Toledo District	17-31 (monthly means)	3000 (mean)	Feb-Apr	20-400	High clay content, stony below 30 cm, overlying limestone, pH near neutral	Ennion (1996)
Fiji	Southern Division	20-30 (monthly means)	2500-3000	Jul-Sep	nda	Deep, reddish clay derived from basic tuff, fertile, pH 3.5-5, sticky and compact, some gley in valley bottoms but generally well-drained.	JICA (1982), DoF (1993), Anise & Moce (pers. comm.)
Guadeloupe	Basse-Terre	23 (annual mean); 13-33 (extremes)	2500-4000	Feb-Apr	50-600	Either permeable, not water retentive, fertility confined to top horizon, fragile, pH 4-5; or water retentive, high in organic matter, deep, fertile	Soubieux (1983)
Honduras	North coast	23-27 (monthly means)	3280 (mean)	Mar-May	40-400	Alluvial, with 30-60% sand, 25-50% loam and 12-19% clay, pH 6-6.5	Portig (1976), Hazlett & Montesinos (1980)
Indonesia	Java	27 (annual mean) 24-31 (monthly means)	1800 (mean)	Jun-Oct	nda	nda	Sukanto (1969)
Martinique	Island-wide	23 (annual mean)	2000-4000	Mar	200-600	Young, rich, deep on moderate slopes, permeable but with a strong water retaining capacity	Portig (1976), Tillier (1995), Chabod (in press)
Mexico	Quintana Roo	25 (annual mean) 6-36 (extremes)	1300-1550	Jan-Apr	10	Thin, limestone-based	Alemán & García (1974), Cuevas et al. (1992)
Philippines	Luzon (Makiling)	27 (annual mean) 40-24 (monthly means)	1820 (mean)	Jan-Apr	110	Volcanic (deep and fertile)	Ponce (1933)
Philippines	Ilocos Sur	27 (annual mean)	2735 (mean)	Dec-Apr	0	nda	Nastor (1957)
Puerto Rico	Luquillo mountains	23 (annual mean)	2300 (mean)	Feb-Apr	130-900	Deep acid clays	Holdridge & Marrero (1940), Portig (1976)
Solomon Is.	Kolombangara	23-31 (monthly means)	3600 (mean)	Oct	0-500	Iron-rich volcanic clay (young and fertile)	Anon. (1979, 1987b)
Sri Lanka	Kurunegala and Kegalle Districts	25-28 (monthly means)	1500-2500	Dec-Mar and Aug	0-500	Red-yellow podzols (well-drained, moderately fine, strongly acid), low humic gleys (poorly drained, fine), red-brown latosols (well-drained, moderately fine, medium acid), immature brown loams (shallow, moderately fine, slightly acid)	Survey Department (1988)

nda = no details available

No other soil experiments are reported and the evidence cited below for different soil preferences is based on largely anecdotal observations in plantations.

Mahogany is not exacting about its soil requirements, growing on most fairly moist sites (Marshall 1939, Chable 1967). In Indonesia, it has been widely reported that mahogany grows well on infertile soils (e.g. Noltee 1926, Japing and Seng 1936, Mijers 1941, Bosbouwproefstation 1949). De Voogd (1948) noted that mahogany grew well on markedly phosphorus deficient soils which had been eroded by shifting cultivation in the Jalappa Forest Reserve, Indonesia. Mahogany has been planted on a wide range of soils in the Solomon Islands (Chaplin 1993) with interesting results. What were considered to be the most fertile sites (moderately weathered and leached, weakly acid to base rich, derived from limestone and other sediments) have not produced significantly faster height growth than the least fertile soils (strongly weathered and acidic, with low concentrations of available and total nutrients). The same does not appear to be true in other countries. Although mahogany may perform better than other plantation species on infertile soils, the species appears to prefer well-drained, fertile soils (Chable 1967, Soerianegara and Lemmens 1994). In Fiji, diameter growth rates of young (under 20 year old) stands on fertile soils are double those on infertile soils (Cottle no date). The volcanic soils of Martinique are very fertile (Tillier 1995) and it is unlikely that the high yields obtained here could have been achieved on poorer soils.

Lamprecht (1989) notes that mahogany can be grown on heavy clays. In Kolombangara, Solomon Islands, mahogany appears to be suited to heavy clays (Anon. 1979) and in Fiji it grows well on the deep and highly weathered red basaltic and andesitic clays (Cottle no date). However, success depends on suitable conditions of soil moisture. Plantations on heavy clays near Siguatepeque in Honduras appear to have failed as a result of excessively dry soil conditions (Chable 1967). Extensive plantings in red clays of Puerto Rico failed due to anaerobic conditions in the sub-soil, which prevented root development (Marrero 1950). Gleying may occur at a depth of 0.5 m or less in high rainfall areas of Luquillo (Francis pers. comm.).

In Fiji the highest growth rates have been measured on alluvial soils (Cottle, no date). Fastest growth in Honduras was found in valleys and littoral areas, which contain alluvial loams (Chable 1967). Holdridge and Marrero (1940) found that the species grows fastest in clay-loams, which tend to be water-retentive but well-drained. Clay-loams or sandy alluvial soil are considered to be the most suitable soils for mahogany in the Philippines (ERDB 1996).

There is disagreement over the suitability of shallow soils for mahogany plantations. Japing and Seng (1936) consider shallow soils to be unsuitable, but in the wet zone of Fiji mahogany is reported to grow well on thin soapstone and alluvial soils (Cottle no date) and in the Philippines planting sites with shallow soils are recommended (ERDB 1996). The apparent contradiction may be a result of various interpretations of 'shallow'. Furthermore, the growth of mahogany will depend on other site variables, particularly soil type and moisture content. For example, Francis (1996) stressed that hybrid mahogany (*Swietenia macrophylla* x *S. mahagoni*) should be planted in alluvial or deep, rich forest soils in dry areas (1000 mm of rain per year), but in sandy or shallow rocky soils in wet areas (1500-2000 mm of rain per year). The same rules will probably hold true for *S. macrophylla*.

Preferred soil pH is reckoned by Webb *et al.* (1984) to range from alkaline to neutral. This is corroborated by observations in Puerto Rico, where mahogany has grown best on limestone hills, particularly in sinkholes where the soil is deep (Tropical Forest Research

Centre 1957, Weaver and Bauer 1986) (see Plate 6.1). Evans (1992) also notes that mahogany grows well on calcareous soils. In Java, mahogany grows best on soils with a pH of 6.5-7.5 (Soerianegara and Lemmens 1994), and Wadsworth (pers. comm.) also believes that a pH neutral soil is probably most appropriate. However, there are also examples of successful establishment on acid soils. In Fiji, mahogany is grown on a large scale in soils which are predominantly of volcanic origin, ranging from pH 5.5 to 4.3 (Busby 1967). Soils with an even lower pH may be satisfactory, but aluminium toxicity may be associated with more acid soils, for example in Puerto Rico (Wadsworth pers. comm.). Soubieux (1983) found that soils in Guadeloupe with a pH below 4.2 and with high aluminium content are associated with low plantation productivity (less than 3 m^3 ha^{-1} yr^{-1}).

Plate 6.1. Mahogany provenance trial in a limestone sinkhole showing trees with rapid growth rate and excellent form. Guajataca Forest, Puerto Rico.

Mahogany appears to grow least satisfactorily on degraded soils. Soils which have been continuously farmed over a number of years have been widely reported to give poor survival and growth rates (Marrero 1942, 1950, Weaver and Bauer 1986, Amador 1988, Francis pers. comm.). This is thought to be because of low contents of organic matter and compact soil structure, which reduces soil water retentivity and renders plants particularly susceptible to desiccation during drought (Holdridge and Marrero 1940, Marrero 1950). Degraded soils may have low vegetation cover and direct insolation may consequently cause soil moisture deficiency during the dry season (Lamb 1966). Conversely, mahogany on sites where a protective overstorey remained intact during cultivation or where second growth forest has been maintained has a much higher survival rate (Marrero 1942). However, with careful site preparation, planting on degraded land may be possible. Perera (1955) describes the establishment of mahogany on degraded lands in Sri Lanka which had been abandoned after intensive food production. Vegetation was burnt and ash and top-soil were scraped into trenches 30 cm deep by 30 cm wide, into which the mahogany seedlings were planted. This method was very successful, giving an average height growth of 3.6 m over a 2.5 year period.

There are a number of other situations where mahogany has been observed to perform poorly. It will not grow on bare laterite, although it can grow in lateritic soils (Troup 1932, Lamprecht 1989). It grows slowly where the soil has been heavily compacted by heavy machinery (Speed pers. comm.). Swampy soils are unproductive (Japing and Seng 1936, Soerianegara and Lemmens 1994, Singh pers. comm.), although mahogany has been noted to survive in poorly drained and seasonally flooded plantations (Chable 1967, Sandom pers. comm.). However, Hummel (1925) was surprised by better than expected growth of a 12-year-old mahogany plantation on a site prone to waterlogging in Belize, and Chinte (1952) reports that mahogany grows well in swamps.

6.3 Topography

Mahogany has been planted on steep slopes to protect steep valley sides from landslides and soil erosion in Indonesia (Fattah 1992), St. Lucia (John pers. comm.), the Philippines (Pablo pers. comm.) and Sri Lanka (Sandom pers. comm.). It is thought that the deep rooting system provides an effective means of stabilising the soil. Run-off erosion experiments under different (unspecified) vegetation types revealed that soil erosion rates under mahogany were the lowest (Coster 1934). The ability of mahogany to colonise steep slopes, accompanied by its tolerance of infertile sites, has led to the establishment of plantations on every feasible slope position within its altitudinal range.

Planting site topography is closely linked to variables of soil depth and drainage. Mahogany reacts to these variables in a similar manner to many other plantation species. Ridges and steep slopes with thin, rapidly draining soils, which induce drought effects, are often associated with high mortality and low growth rates in mahogany plantations (Weaver and Bauer 1986, Neil 1986, Lamb 1955, 1966). Lower slopes with gentle gradient and well-drained valley floors tend to produce the fastest growing trees (Weaver and Bauer 1986, Soubieux 1983, Cree 1953).

Although soil in valley bottoms is likely to be deeper and more fertile than on steep slopes and ridges, a high clay content may create conditions of poor drainage and susceptibility to waterlogging. Poor growth in valley bottoms has been recorded in Vanuatu and the Solomon Islands (Kolombangara) (Neil 1986, Lilo pers. comm.). In Puerto Rico, trees on bottomland sites of a 1974 planting of hybrid mahogany (*Swietenia macrophylla* x *S. mahagoni*) displayed similar survival to those on ridges and significantly lower survival than those on intermediate slopes (Weaver and Bauer 1988). Marrero (1950) noted that in plantations on red clay sites (which were abandoned), only trees on the lower concave slopes would have produced timber. Bauer (1987a) reports that height growth rate in Puerto Rico is lowest on valley bottoms and highest for mahogany on slopes over 40%, particularly on concave slopes. In soils prone to waterlogging the preferred planting site would therefore appear to be somewhere on the lower to mid-slope. If soils are deep enough, even ridge sites may be suitable (Francis pers. comm.).

6.4 Indicative species

Species which act as indicators of appropriate planting sites for mahogany are poorly described in the literature. Holdridge and Marrero (1940) note that site factors which favour mahogany closely match those which favour coffee, although mahogany is tolerant of a minimum annual rainfall less than the 1200 mm required for coffee (see Section 6.1). In the Caribbean, *Buchanavia tetrophylla* and *Manilkara bidentata* are good general indicators (Francis pers. comm.). Blue mahoe (*Hibiscus elatus*) has a narrower range of site preferences which coincide closely with those of mahogany.

For planting sites within the natural range of mahogany, readers are referred to Lamb (1966), who describes natural mahogany associations in considerable detail. However, Lamb does not identify indicator species for suitable mahogany plantation sites.

6.5 Summary

- Mean monthly temperatures in mahogany planting sites should fall within the 15-35°C range. Suitable rainfall figures depend on local evapotranspiration rates. Mahogany has grown well on sites receiving rainfall of 1000-4000 mm yr^{-1}. A dry season of up to 4 months can be tolerated.
- Mahogany appears to grow satisfactorily on a wide range of soils, from clays (but not gleys) to coarse sandy soils. Slightly alkaline to pH-neutral soils are preferred, but acid soils with a pH as low as 4.5 give acceptable results. Generally mahogany seems to prefer well-drained soils, although in dry climates, more water-retentive soils are preferable. Plantations can survive periodic waterlogging.
- The highest growth rates in unfertilised plantations have been recorded on volcanic soils of Martinique. Mahogany appears to be able to tolerate nutrient deficiencies which some other timber species, such as teak, cannot. Overcultivated soils degraded of organic matter, compacted soils and laterites give lowest growth rates.
- Mahogany can be successfully established on steep and unstable slopes and is effective in reducing soil erosion.
- Stand productivity is affected by the interaction between patterns of rainfall, soil type and slope position. Where soils are prone to waterlogging, mid- or high-slope positions give most rapid growth rates. In areas of high rainfall, shallow soils on ridges may be suitable planting sites.
- Few species indicative of suitable mahogany planting sites have been identified; coffee is the only well-known example.
- Current information on the site preferences of mahogany is largely anecdotal. Further investigation of the site preferences of mahogany is strongly recommended for those interested in improving plantation yield.

Chapter Seven
Plantation Establishment

Over the last century mahogany has been planted on a wide range of sites including cleared ground, grasslands, secondary forest, disturbed natural forest and even relatively undisturbed natural forest. Many believe that the maintenance and manipulation of some or all of the pre-existing vegetation can be used to reduce shoot borer damage (see Section 11.1.2.3). The decision to retain natural vegetation will affect decisions on site preparation, stocking density, species composition and the overall layout of the plantation. Associated natural vegetation may affect the growth and development of mahogany quite significantly.

Examples of mahogany plantation establishment from around the world are therefore categorised here according to vegetation retained at the time of planting. A simple distinction has been made between planting 'in the open', defined as sites which have no natural woody vegetation present (see Section 7.2) and planting 'in natural forest', defined as sites which do have natural woody vegetation present (see Section 7.3). Information contained in the literature lacks the necessary detail (species composition, forest profile, canopy height and density) for further subdivision.

The majority of mahogany plantations have been established using nursery stock, but direct seeding is a possible alternative. Direct seeding may be used in open or natural forest sites. However, the technique has been described separately (see Section 7.1) to allow direct comparison of many different experiences.

7.1 Direct seeding

A number of direct seeding trials have been reported. In Vanuatu, 63% of seeds sown in pairs at a depth of 7.5 cm germinated after 1 month (Leslie 1994). In contrast, direct seeding experiments carried out by Perera (1955) gave 24% survival. However, mean height after nearly 2 years was 2.0 m (ranging from 0.9 to 2.4 m) which compared favourably with the mean height of 2.3 m for planted striplings over the same period (range: 0.6-3.0 m). Lamb (1955) also found little difference in growth between direct seeded and planted seedlings after 1 year. Negreros (1991) carried out an experiment to see if germination rates were affected by the method of sowing. Mahogany seeds were either scattered (3400 seeds) or dibbled (4100 seeds) under a natural forest overstorey. There was no significant difference between the two treatments in terms of germination rate (41% vs 47%), survival of germinated seedlings (79% vs 75%) or mean height (22 cm vs 20 cm) after 12 months. Seed viability was 95% (estimated) before the experiment took place.

Direct seeding was first used for the establishment of mahogany on a large scale in Belize. In order to avoid the costs associated with rearing and planting mahogany seedlings in the nursery, seeds were sown in a taungya system which took care of planting and weeding costs for the first year (see Section 7.2.2.1). The practice in Belize was to sow two seeds per peg (Lamb 1947). Where both seeds failed to germinate, seedlings were transferred from other pegs where both seeds had survived. Thus a survival of 50% gave 100% of the desired stocking after transplanting. In fact a survival as low as 20% was considered acceptable. Since the desired stocking density at the end of a rotation was 125 trees ha^{-1}, the standard spacing of 4.5 x 4.5 m (Cree 1953) gave this stocking figure with a survival rate of 12.7%. However, with the combined threat of shoot borer and hurricane damage in Belize, higher stocking levels would have been necessary to ensure a final crop of reasonable quality.

The success of the direct seeding method in terms of germination and survival of mahogany seedlings in Belize is documented in Forest Department Annual Reports (see Table 7.1). The terms 'germination' and 'survival' are not clearly defined in the literature and may have been used interchangeably. Not all plantations were assessed. Poor survival and complete failure were common. It appears that low germination rates were the result of unsuitable conditions of rainfall after sowing and low seed viability. No mention is made of unsatisfactory weeding and it appears that plantations were well maintained by farmers over the first year.

Table 7.1. Germination and survival of direct seeded mahogany in Belize

Plantation name	Size (ha)	Germ. (%)	Surv. (%)[a]	Explanation / notes	Source
nda	'large areas'	failed		non-viable seed (stored too long)	Nelson Smith (1942)
Silk Grass	nda	40	-	heavy rains followed planting	Nelson Smith (1942)
Silk Grass	nda	15	-	long dry period after planting	Stevenson (1943)
Silk Grass	4	-	77	nda	Stevenson (1944a)
Silk Grass	nda	-	95	nda	Lamb (1946)
Iguana Creek	4	-	66	seed sown soon after collection	Lamb (1946)
Silk Grass	4	25	-	seed from distant source (Cayo)	Lamb (1947)
Silk Grass	nda	60	-	seed from local source	Lamb (1947)
(Toledo District)	4	-	88	nda	Lamb (1948)
Silk Grass	16	21	-	nda	Lamb (1949)
Commerce Bight	16	22	-	season 'abnormal'	de Silva (1952)
Topco Estate	8	-	57[b]	nda	de Silva (1952)
Grant Works	8	-	58	nda	Cree (1953)
Silk Grass	8	failed		dry weather after planting	Cree (1954)
Freshwater Creek	2	failed		dry weather after planting	Cree (1954)
Columbia River	23	-	20	nda	Cree (1956)
Columbia River	78	-	36	nda	Cree (1957)
Silk Grass	8	40	-	nda	Cree (1957)
Columbia River	102	10	-	nda	Frith (1962)

[a] Measurements taken after about a year, except: [b] measurements taken after 2-3 years
nda = no details available

The use of nursery planting stock was tried on a small scale in Belize, but it seems that forest managers were deterred by failure (Stevenson 1944b, Lamb 1949). At the time it was suggested that the long tap root and lack of side roots meant that mahogany seedlings were unsuitable for transplanting (Lamb 1949), although others did manage to use nursery plants successfully in Belize (e.g. Cree 1953).

Recent direct seeding trials in Fiji have given very positive results (SRD 1995). Despite years of successful out-planting of nursery stock, forest managers now believe that direct seeding improves seedling survival (Singh pers. comm.). Roots are thought to develop significantly better when left undisturbed. A new policy of 50% planting and 50% direct sowing has therefore been put into effect in the Southern Division (Seroma pers. comm.). A mound is dug and 3 seeds are sown at a depth of less than 2.5 cm. Seeds sown at a greater depth have been found to decay. Of the seeds, 50% germinate within 3 weeks but germination continues over a period of 60 days. Seedlings are checked at 3 months and, if beating up is required, seedlings are transplanted from one mound to another. Extra seedlings are kept for 6-9 months in case they are needed to fill gaps, then pulled up. So far there have been no problems with predators of mahogany seed or the young seedlings. Survival after direct sowing has been found to be approximately 85% (Evo pers. comm.). Seeds have been found to be resistant to drought after sowing, but periods of prolonged heavy rain cause the seeds to rot (Singh pers. comm.). Trees shading the planting lines must be poisoned before sowing in wetter areas in order to let the sun dry the soil sufficiently to prevent the seeds from rotting.

Direct seeding was used to establish the 1963 mahogany plantations in the Luquillo Forest of Puerto Rico (Weaver and Bauer 1983, 1986). The spacing was 3 x 3 m, with 3 seeds per hole (presumably due to doubts over seed viability). Lamb (1966) reports the use of 4 seeds per hole in other Puerto Rican plantations, finding that nearly every hole had produced at least one living seedling 30-45 cm high after the first dry season. However, Streets (1962) reports that direct seeding in Malaysia was not successful.

Direct seeding has also been used to increase stocking densities of mahogany in logged natural forest. At Freshwater Creek in Belize, 329,000 mahogany seeds were sown in old extraction routes in areas where mature mahogany had been extracted (Frith 1959). Stanley (pers. comm.) reports the use of the technique in the Petén, Guatemala, to enrich natural mahogany stands.

It is difficult to determine from these experiences exactly which site conditions make direct seeding an appropriate method to use. From observations in Fiji (see above) it appears that well-drained soils are most suitable, although much will depend on local climate. Lamb (1966) suggests that the technique is best-suited to drier planting sites in tropical dry forest (defined as having a mean annual rainfall of 1780 mm and a long dry season). In wetter areas, high quality nursery stock gives more vigorous early growth and is better able to compete with weeds than freshly germinated seeds. However, under a taungya or other agroforestry system, where regular weeding of mahogany seedlings can be carried out as part of normal maintenance of agricultural crops, direct seeding is more likely to be successful under a much wider range of site conditions.

7.2 Planting in the open

A light-demanding species such as mahogany will grow most rapidly on open sites, although some information sources associate conditions of high light availability with increased incidence of shoot borer attack (see Section 11.1.2.3). In fact, there are grades of 'open-ness' within the definition of open sites. If left, tall weeds and grasses could potentially obscure mahogany seedlings from the shoot borer. However, such advantages would have to be balanced against the reduction in growth rate likely to result from increased competition from weeds.

In this section, a distinction has been made between mahogany plantations that are pure and those that are mixed. This distinction is valuable when describing the establishment of mahogany. However, by the time the mahogany trees reach maturity, the composition of an initial mixture may have changed. In some cases, the proportion of mahogany trees in the plantation will decline. For example, in Java *Toona sureni* and occasionally teak may grow up through mahogany plantations and these species are often favoured by local foresters over the mahogany (Matsumoto pers. comm.). Where it is growing vigorously, the proportion of mahogany in the plantation may increase with time. For example, in Sri Lanka mixed stands of mahogany and *Artocarpus integrifolia*, initially thinned to a ratio of less than 2:1 (Forest Department archives), became dominated by mahogany within a period of 20 years (Perera 1955). The distinction between pure and mixed is not, therefore, meant to indicate a permanent state.

7.2.1 Pure mahogany plantations

Pure mahogany plantations have been established in a number of countries (see Table 7.2). Young plants are usually set out at a spacing of 2-3 m. In general, planting sites are cleared of woody growth, but weedy regrowth or tall grass are often present and may need to be cleared before seedlings can be planted. The use of close spacing has the associated advantage of early canopy closure, which shades the ground layer and reduces weed growth between trees. Close spacing will also serve to limit development of lateral branches.

The results of pure plantings are varied and their success depends largely on local shoot borer populations (see Table 7.2). In places where shoot borer attack does not appear to be a major problem (e.g. Java and the Philippines), pure plantations have been very successful. Where shoot borer attack is severe (e.g. Malaysia and Trinidad), pure plantings have given poor results. There are no large-scale plantations in Mexico and the multitude of small, privately owned mahogany woodlots have produced very variable results. However, it is worth noting that despite considerable shoot borer damage to pure plantations in Martinique, problems of poor form are overcome by selective thinning and pruning (Chabod in press). There is some evidence that densely stocked stands will recover form, given time (see Section 11.1.3.4).

Table 7.2. Techniques used to establish pure plantations of mahogany

Spacing (m)	Site preparation	Results	Country	Source
1.8x1.8 or 2.4x2.4	nda	Used 1931-63; successful where matched to site	Puerto Rico	Bauer (1987a)
2.0x2.0	Pre-existing vegetation cleared	Successful despite shoot borer attack	Martinique	Chabod (in press)
2.0x2.0	In *Imperata* grassland; 1 m diam. circles weeded	Successful; little shoot borer attack	Philippines	Chinte (1952), San Buenaventura (1958)
2.0x2.0 or 3.0x3.0	Strips 1 m wide weeded	Close spacing better where weeds are a problem	St. Lucia	James (pers. comm.)
2.0x2.0 or 3.0x3.0	1 m diam. circles or 1.5 m wide strips weeded	nda	Venezuela	Bascopé *et al.* (1957)
2.5x2.5 or 3.0x3.0	Pre-existing vegetation was cleared	Invasion by herbaceous vegetation killed many plants	Guadeloupe	Soubieux (1983)
2.0x4.0 or 3.0x3.0	Pre-existing vegetation was cleared	Successful on a large scale; little shoot borer attack	Indonesia (Java)	Matsumoto (pers. comm.)
3.0x3.0	nda	Unsuccessful: severe shoot borer attack	Malaysia (Sabah)	Kotulai (pers. comm.)
3.0x3.0	1 m diam. circles or strips weeded	Successful in some locations	Mexico	Flinta (1960), Cuevas *et al.* (1992)
3.0x3.0	1 m diam. circles or 1.5 m wide strips weeded; a 30 cm trench dug around each plant	nda	Martinique	Marie (1949)
3.7x3.7	nda	Trees heavily branched	Trinidad	Streets (1962)
5.0x5.0	Pre-existing vegetation clear-felled and burnt	nda	Malaysia (Sarawak)	Chai & Kendawang (1984)
nda	nda	Reasonable early growth; shoot borer attack caused branching after 2-3 years	Malaysia (Peninsular)	Streets (1962)

nda = no details available

7.2.2 Mixed mahogany plantations

The establishment of mixed plantations has been recommended primarily because there is some evidence that mahogany in mixtures is less susceptible to shoot borer attack (see Section 11.1.2.3). Furthermore, mahogany is a relatively long-rotation species and the inclusion of faster-growing species in the plantation may bring early financial returns. The use of mixtures does not necessarily entail reduced stocking or yield of mahogany: agroforestry systems can make use of otherwise unused land between mahogany seedlings before canopy closure.

7.2.2.1 Agroforestry systems

The size and density of the canopy of a mature mahogany crown mean that the species is often not considered to be an appropriate component of long-term agroforestry systems. Systems therefore tend to be designed specifically for the establishment of mahogany, with agricultural crop species chosen to assist this goal. However, mahogany may be successfully combined with perennial food crops such as coffee at low densities; trees may be pruned to reduce the extent of shade and improve form.

The use of mahogany in agroforestry systems is not a recent phenomenon. The taungya system was used to establish mahogany on a large scale during the 1940s and 1950s in Belize (see Box 7.1) and in many Caribbean countries (see Table 7.3). One of the reasons for the adoption of the system in the Caribbean was the shortage of farming land caused by private ownership of large estates of monocultures (Huguet and Marie 1951). The agricultural crops most commonly used were maize and bananas. The system was often successful because the plantations were well managed over the first 2 years or while crop production is maintained (Soubieux 1983), which is a critical period for the young mahogany.

Box 7.1. Mahogany taungyas in Belize

During the 1920s and 1930s, officers of the Forest Department of British Honduras (now Belize) focused on developing techniques for increasing the stocking density of mahogany in natural forest. However, by the mid-1940s (following the Depression and a number of devastating hurricanes) attention had shifted to the establishment of mahogany plantations, which were considered to have greater productive potential. The taungya system was thought to be the most appropriate method of establishment. It was estimated that planting mahogany at a rate of 60 ha yr^{-1} would be sufficient to replace all the mahogany being cut from Crown land in the colony, assuming a final crop of 125 trees ha^{-1} (Lamb 1948). Planting targets were increased in later years to 100 ha yr^{-1} (Cree 1955), but rarely achieved owing to shortage of seed, low survival of seedlings and insufficient man-power. The busiest planting period was between 1955 and 1964, when 591 ha were established in Columbia River Forest Reserve, Toledo District alone (Ennion 1996).

District	Period of establishment	Area planted (ha)	Managed area remaining by 1964
Cayo	1945-54	119	0
Northern	1953-63	40	40
Stann Creek	1930*-61	120	70
Toledo	1928-63	617	519
Total	-	896	629

* Date uncertain   Source: Forest Department Annual Reports 1945-65

Taungya plantings were carried out in logged natural forest or secondary regrowth (Lamb 1948). The understorey of shrubs and small trees were felled (Cree 1955). Many canopy trees were removed as well, but a number of tall nurse trees were retained to provide a dappled shade (Nelson Smith 1942, Cree 1955). The shady conditions were thought to reduce shoot borer damage, but also reduced height growth rates by around 40% (Nelson Smith 1942).

Although some small nurseries produced seedlings for planting, virtually all the mahogany plantations were established by direct seeding. Mahogany was sown at a range of spacings, which were sometimes based on availability of seed rather than the type of taungya. 3 x 3 m was most commonly reported (Nelson Smith 1942, Stevenson 1944b), but variations up to 6 x 6 m were recorded (Lamb 1947). By 1952 the standard spacing for mahogany was 4.5 x 4.5 m (Cree 1953). However, spacing ultimately depended on the success of seed germination and seedling survival, which was very variable. The use of 'Anderson groups' spaced at 9 m intervals was tried in 1956 (Cree 1957). Seeds in the group were sown at a spacing of 20 x 45 cm. The number of trees growing in each group depended on germination and survival with a maximum of 14, although 6 or more was considered sufficient. During the last few years of plantation establishment, promising results were obtained (Frith 1959). Groups were preferred because they gave a better selection of evenly distributed 'elite' trees for the final crop.

Mahogany was usually sown amongst a crop of maize, but plantains and cassava were also used (Lamb 1948). A mixture of crops were often planted, e.g. maize, sesame and plantain (De Silva 1952). Creole rice was planted at one stage, due to a threatened shortage, but was not thought

> **Box 7.1. (continued)**
>
> to be an appropriate species to be planted with mahogany (Lamb 1947). Crops were sown a month before the mahogany (De Silva 1952). The area was abandoned (by farmers) after the first maize crop was harvested (Kinloch 1933). Mahogany was then considered to be able to cope with weed growth, which was allowed to overtop it to reduce borer attack.
>
> Amerindians were paid for demarcating farms and for sowing the mahogany seed. There was usually good co-operation with the Forest Department (Lamb 1948). However, in some areas there was an insufficient number of labourers and Forest Department staff carried out the work themselves (Cree 1953). The most successful area was the Columbia River Forest Reserve, where the Forest Department licensed an area of land to the local Amerindians annually (Ennion 1996). In this area maize crops were sometimes spectacular (Frith 1958) which encouraged participation in the following year (Frith 1959). The system covered the costs of clearing the land, sowing with mahogany and tending for one year (Lamb 1947).
>
> The plantations established by taungya were not always pure mahogany. *Cedrela odorata* was also used (Ennion 1996), often for beating up or when mahogany seed was scarce. Other species were mixed with mahogany and Cree (1956) records the establishment of 12 ha of mahogany mixed with teak, *Gmelina* and mayflower (*Tabebuia pentaphylla*) sown in maize at Silk Grass. Natural regeneration of certain marketable species was sometimes encouraged: in Chiquibul the ground between groups of mahogany was hand-weeded to encourage prolific balsa regeneration after the agricultural crops had finished (Cree 1957).
>
> High pruning was carried out in some plantations (Stevenson 1944a, Lamb 1949). Cleaning of young taungya was found to be necessary twice a year (Lamb 1949). Thinning of natural species (not mahogany) was also practised, but stocking densities before and after treatment were not recorded. The term 'thinning' seems to refer to two separate techniques:
> 1. A liberation thinning that removed naturally regenerated trees and poorly formed mahogany from the vicinity of a selected mahogany tree. The stocking was very variable so a systematic thinning would have been inappropriate.
> 2. A nurse tree thinning that removed remaining canopy trees. This was carried out at about 6 years from sowing, leaving mahogany crowns exposed to full light (Lamb 1948, Frith 1962).
>
> Response to thinning led to a doubling of annual dbh increment of dominant mahogany trees (Lamb 1949). In the most successful plantations a closed canopy was formed in about 10 years and boles were reported to clean up successfully after earlier shoot borer attack (Stevenson 1939).
>
> During the 1950s, most plantations appear to have been cleaned and thinned regularly (Cree 1953-57). Growth rates were initially high: on one plot after three years, an average diameter of 1.8 cm and height of 4.5-6 m was recorded (Lamb 1947). However, overall, the quality of plantation maintenance was very variable. Some of the earlier plantations were neglected. For example, in the Columbia River Forest Reserve certain areas were found to be in poor condition, due to second growth cover of polak (*Ochroma limonensis*) and Moho (*Heliocarpus donnel*), or thick choking creepers (Stevenson 1939). Many plantations failed or were abandoned owing to low survival.
>
> After 1964 the plantations received no further maintenance (Ennion 1996). Of the main plantation area in the Columbia River Forest Reserve, only 200 ha remained by 1981. This was further reduced by milpa farmers to 104 ha by 1985, and another area has since been cleared for farming (Ennion 1996). It is fortunate that 4 permanent sample plots have survived in this area. Measurements have been taken at irregular intervals over a period of 38 years (see Section 9.1 for more details). Despite the lack of maintenance, growth rates are surprisingly high, with a mean diameter increment of nearly 1 cm yr^{-1} (Ennion 1996).

A number of research trials involving the use of mahogany in agroforestry systems have been reported recently (see Table 7.4). Research in Brazil indicates that an agroforestry system which uses a mixture of tree species may limit shoot borer damage (see Section 11.1.1 for possible mechanisms). Taungya trials in the Solomon Islands (see Plate 7.1) and Trinidad (see Plate 7.2) have produced good results, but offer little protection from the shoot borer. Trials indicate that mahogany performs well in comparison with the other tree species used in agroforestry systems.

Table 7.3. Agroforestry systems used to establish mahogany plantations (see Box 7.1 for details of experience in Belize)

Country (region)	Agricultural crops	Spacing	Establishment and maintenance	Source
Indonesia (Java)	Maize, tapioca, upland rice, cassava	nda	nda	Lamprecht (1989) Soerianegara & Lemmens (1994)
Martinique	Various food crops	Mahogany at 3x3 m; or 4x4 m up to 2x2 m after cropping	Forest clear-cut and stumps removed; fertiliser added where bananas used; best results for mahogany with bananas	Marie (1949), Huguet & Marie (1951)
Puerto Rico	Maize, beans, plantains, cassava, bananas	nda	Bananas used to provide shade for mahogany	Marrero (1950), Weaver (1989)
St. Lucia	Bananas	Bananas at 3x3 m; mahogany interplanted	Bananas maintained for up to 10 years; farmers responsible for bananas and weeding	Vidal (pers. comm.)
Trinidad	Bananas, coffee	nda	Mahogany established under bananas with *Cedrela odorata*, *Erythrina* spp., *Cordia alliodora*; rotation length 30-40 years; later underplanted with coffee	Lamb (1969)

nda = no details available

Plate 7.1. A trial mahogany-maize taungya system. The trees have been pruned to maintain form after shoot borer attack. Kolombangara, Solomon Islands.

Plate 7.2. A trial mahogany-banana taungya system. Trinidad, West Indies.

The length of time over which crop production can be maintained will vary according to spacing of the tree species. With rows of mahogany less than 3 m apart, the canopy becomes sufficiently dense after 2 years to prevent intercropping with maize (see Table 7.4), unless the trees are regularly pruned. This conclusion has been confirmed by observations elsewhere (e.g. Anon. 1959). However, the use of relatively shade-tolerant agricultural species, such as cocoa and coffee, may allow the period of agricultural crop production to be extended.

Mahogany has been used as a shade tree for cocoa on private estates in Belize (Hummel 1925), the Philippines (ERDB 1996) and Trinidad (Streets 1962). In a trial on 2 x 0.4 ha plots in Belize, the first cocoa crop was reaped 3 years after planting, following careful manipulation of the mahogany canopy to create suitable light conditions (Frith 1964). However, by the end of the 1960s, cocoa and coffee was being grown under open conditions with fertiliser (Lamb 1969) and there have been few reports of the use of mahogany as a shade tree since then.

Because of their similar site requirements, declining cocoa or coffee plantations may provide appropriate conditions for the establishment of mahogany. Lamb (1955) describes the conversion of cocoa plantations to mahogany in Trinidad. Trees were planted in groups, the size of which depended on gaps in the cocoa, at 15 x 15 m. This spacing was considered sufficient to form a closed mahogany canopy when trees eventually reached maturity.

Table 7.4. Trial agroforestry systems incorporating mahogany

Country (region)	Agricultural crops	Non-mahogany tree species	Spacing	Results	Source
Brazil (Central Amazon)	Maize, banana (planted together)	*Cordia goeldiana*	nda	Both tree species grew satisfactorily over 8 years. The system could be improved by introducing fruit trees for mid-rotation productivity	Brienza & Yared (1991)
Brazil (Central Amazon)	Maize, banana (planted together)	*Dypterix odorata*, *Vochysia maxima*, *Bertholettia excelsa*, *Inga* spp., *Cordia goeldiana*, *Bagassa guianensis*, *Theobroma grandiflorum*	nda	Mahogany showed 100% survival and better height growth than all other tree species	Brienza & Yared (1991)
Brazil (Para)	Maize	None	Mahogany at 7.5x7.5 m	Costs of planting were covered by income from maize. Mahogany growth in 1 year was impressive (2.5-3.0 m)	Verissimo et al. (1995)
Mexico	Maize	*Cedrela odorata*, *Cordia alliodora*	Trees at 2.8x2.8 m or 2.0x2.0 m	Maize was grown for 2 years after which 6 weedings were carried out before canopy closure. Spacing has no effect on growth rates over a 7 year period. Mahogany survived best out of the 3 spp. used (80%)	Poras & Luyano (1974)
Solomon Islands (Kolombangara)	Banana, sweet potato, maize, beans (planted in rotation)	None	Mahogany at 5.0x2.5m	The trial is ongoing, with the intention of maintaining crop production for 5-8 years. Crop productivity has been average. By year 3 mahogany had developed bushy crowns due to shoot borer attack	Isles & Viuru (pers. comm.)
Vanuatu	*Sorghum* sp., *Desmodium intortum*, *Panicum maximum*, *Setaria sphaceleta* (planted separately)	None	nda	After 2 years, there was no significant difference in mahogany volume between the different mixtures. *D. intortum* was least effective at hurricane protection	Leslie (1994)

nda = no details available

Table 7.5. Mahogany plantations containing a mixture of species

Country	Non-mahogany spp. in the mixture	Spatial relationship between spp. in the mixture	Results	Source
India (Malabar)	*Tectona grandis* (teak)	Mahogany planted irregularly in 30-40-year-old teak	Mahogany grew well	Streets (1962)
India (Madras)	*Trema orientalis*	Mahogany planted with or in *T. orientalis*	The mixture gave good protection from the shoot borer and the collar borer, *Pagiophloeus longiclavis*	Streets (1962)
Indonesia	*Cassia simea* and *Leucaena glauca*	Alternate rows of mahogany/ *C. simea* mix and *L. glauca*	nda	Noltee (1926)
Indonesia	*Dalbergia latifolia* and *Eucalyptus* sp.	Overall spacing 4.0x4.0 m	The mixture gave good results on drained, previously marshy sites	Niebroek (1940)
Indonesia	*Eucalyptus platyphylla*	nda	nda	Coster & Van den Blink (1939)
Indonesia	*Peltophorum pterocarpa*	Mahogany spacing 6.0x6.0 m; *P. pterocarpa* in mahogany	*P. pterocarpa* prevented the invasion of weeds and produced tan bark	Spaan (1909)
Indonesia	*Pinus merkusii*	nda	The mixture was successful only where intensively tended	Winanto (1958)
Indonesia	*Tectona grandis* (teak)	Mahogany planted in teak	Mahogany grew well in dense teak plantation but had an adverse affect on teak development	Van Hall (1922), Becking (1928)
Malaysia (Peninsular)	*Albizia flacata*	Mahogany planted in *A. flacata*	nda	Streets (1962)
St. Lucia	*Hibiscus elatus**	*H. elatus* planted in young mahogany; overall spacing less than 2.0x2.0 m	*H. elatus* grew slightly faster than mahogany, becoming dominant. The stand has recently been thinned to favour the mahogany. Although suppressed, the form of mahogany trees is very good	Bobb (pers. comm.)
Sri Lanka	*Artocarpus integrifolia* (jak)*	Mahogany planted at about 5.0x5.0 m in quincunx with 3-10-year-old jak at 2.5x2.5 m	The mahogany grew vigorously, eventually suppressing the jak. There was good recovery from early shoot borer attack	McNeil (1935), Perera (1955)
Trinidad	*Tabebuia pentaphylla* and *Gliricidia maculata*	*G. maculata* planted last to fill the gaps	nda	Streets (1962)

* Mixed plantations known to exist today

nda = no details available

Table 7.6. Mixtures with mahogany in research plots and trials

Non-mahogany spp. in mixture	Spatial relationship between spp. in the mixture	Results	Country (region)	Source
Cordia alliodora	C. alliodora planted in mahogany at 5x5 m	C. alliodora harvest planned after 15-20 years	Vanuatu	Neil (1986)
Gmelina arborea	Mahogany planted in 3-year-old G. arborea	nda	Malaysia (Sarawak)	Butt & Chiew (1982)
Leucaena leucocephala	Overall spacing 5x3 m	The form of mahogany was above average after 9 years. However, L. leucocephala is susceptibile to the aggressive fungus Phellinus noxius and the psyllid, Heteropsylla cubana	Solomon Islands	Anon. (1988a)
L. leucocephala	L. leucocephala planted at 4x1 m	Mahogany planted with nurse crop had a higher survival rate than without. L. leucocephala thinning planned at year 4 to release mahogany	Philippines	Quintana (1986)
L. leucocephala	nda	Nitrogen fixing properties of L. leucocephala failed to enhance performance of mahogany. Soils (acid or P-deficient) unsuitable for L. leucocephala	Philippines	Granert & Cadampog (1980)
L. leucocephala	Overall spacing 5x3 m	After 6-7 years, the form of 6.3 ha of mahogany was above average	Fiji	Wilson (1986)
L. leucocephala and A.* falcataria	Mahogany planted in L. leucocephala and A. falcataria	nda	Indonesia	Soerianegara & Lemmens (1994)
Pinus caribaea	Mahogany planted in P. caribaea	After 8 years, mahogany was found to have grown well (average dbh of 9 cm) despite only light of thinning of pine. Suggested as a means of raising the value of poor pine plantations	Sri Lanka	Weerawardane (1996)
Securinega flexuosa	Mahogany planted at 8x6 m in 2-year-old S. flexuosa; overall spacing 4x3 m	After 3 years, mahogany had a greater length of unbranched stem (averaging 7-8m) than the pure mahogany control	Solomon Islands	Anon. (1987b, 1988a), (Iputu pers.comm.)
Terminalia calamansanai	Mahogany planted at 10x3 m in rows with T. calamansanai; overall spacing 5x3 m	Rapid height growth and large spreading crown of T. calamansanai indicated that it was not an appropriate species for use with mahogany	Solomon Islands	Anon. (1987b, 1988a)

* Albizia (=Paraserianthes)   nda = no details available

7.2.2.2 Forestry systems

Many mixed forestry plantations have been established with mahogany as a component (see Table 7.5). In general, mahogany has been planted either in a pre-existing plantation of timber species before canopy closure (e.g. *Artocarpus integrifolia* or *Tectona grandis*), or on a clear site with a fast-growing species (e.g. *Leucaena* or *Eucalyptus* spp.). The latter technique creates similar light conditions to the former after a few years, i.e. lateral rather than overhead shade. The use of the term 'underplanting' (of a plantation species with mahogany) is common in the literature, but perhaps a little misleading. The term correctly implies a difference in height between the species in the mixture, but also implies that mahogany is planted directly under the canopy of the non-mahogany species, which is not usually the case. The technique used could perhaps be better described as 'inter-planting', usually in rows.

In the long term, a mixture is defined primarily by the relative crown positions of its constituent species. Wormald (1992) describes a simple two-way classification:
- single-layered canopy: the mixture is generally permanent, giving a multi-specific end result;
- two-layered canopy: the mixture may be permanent, giving a multi-specific end result, or temporary, giving a mono-specific end result.

Most examples of mixed mahogany plantations (see Table 7.5) fall into the latter category. The primary function of such mixtures is to increase the short-term profitability of the plantation, achieved by planting a faster-growing, shorter rotation species with the mahogany. The final objective is usually the creation of a mono-specific mahogany stand (because of the high value of mahogany timber) and the mixture is therefore temporary. The main difficulty is the extraction of the shorter rotation length species without damaging the mahogany (Wormald 1992). The more intimately the species are planted, the more care and expense is required.

The best documented example of an attempt to produce a permanent mixture with a single-layered canopy comes from Sri Lanka, where approximately 3000 ha of mixed mahogany and jak (*Artocarpus integrifolia*) plantations were established (Sandom and Thayaparan 1995). Mahogany was planted in quincunx with jak when the latter was 3-10 years old. In addition to increasing the value of the final crop, it appears that the purpose of planting mahogany was to prevent a grassy sward from developing under pure jak plantations. The canopy was too open and the grass hampered growth of the young trees (Anon. 1935). The light conditions proved to be ideal for mahogany. Notes which accompany permanent sample plot information report that, despite considerable shoot borer attack, the mahogany recovered well after a few years (Forest Department archives, unpublished). Efforts were made by forest managers to liberate sub-dominant jak crowns (Perera 1955), but mahogany became the dominant species and many plantations were almost pure 30 years from the time of establishment (Anon. 1959). The mixture was very successful from the mahogany perspective, but the jak was clearly unable to compete effectively. Wormald (1992) notes that management of such mixtures in a single canopy is extremely difficult and is possible only on a narrow range of sites if both species are to be retained.

The possibility that the use of mixtures may reduce damage caused by the shoot borer has led recently to the establishment of a number of research plots and trials (see Table 7.6). In most cases, pure mahogany control plots have not been established and it is therefore impossible to quantify the protection that the mixtures may afford. However, a thorough investigation into the use of mixtures has been carried out in the Solomon Islands with interesting results (see Box 7.2). In general, mixtures which permit mahogany to grow vigorously have had some success in limiting shoot borer damage (see also Section 11.1.2.3). Favourable results have prompted other authors to suggest alternative species for mixing with mahogany. These include: *Acacia auriculiformis* for Sri Lanka (Sandom and Thayaparan 1994) and the Solomon Islands (Anon. 1988a); *Acacia mangium* for the Solomon Islands (Anon. 1988a); *Albizia* (=*Paraserianthes*) *falcataria* for the Solomon Islands (Jones 1976); *Eucalyptus deglupta* for the Philippines (Lapis 1995); *Gmelina arborea* for the Philippines (Malvar pers. comm.) and for difficult sites in Sri Lanka (Sandom and Thayaparan 1994).

Before establishing a mixed plantation, forest managers need clear objectives, experience and the capacity for continuity of management (Wormald 1992). The success of the mixture will be determined not only by the ecological requirements of the species concerned, but also by the silvicultural treatments that are carried out. Mixed mahogany stands require careful maintenance to ensure that the mahogany receives sufficient light, and the timing of treatments is critical to the health and productivity of the stand. Labour-intensive operations may be required, but evidence indicates that properly managed mixed stands produce well-formed mahogany trees.

7.2.2.3 Silvipastoral systems

There has been some interest in the possibility of grazing cattle under mahogany plantations in Western Samoa (Pottier 1984). Because of the compact crown of young trees, light transmission through low density (300 stems ha^{-1}), 6-year-old plantations was found to be as high as 70-75%, enabling pastures to maintain their productivity. In a cattle grazing experiment, pastures under mahogany plantations were found to support cattle for 112 days ha^{-1} yr^{-1}. Pottier estimates that a mahogany silvipastoral system could support cattle for up to 10 years. One drawback to the system was the development of a mahogany-bark appetite by cattle approximately one month before the wet season. Once started, the bark eating habit continued all year round. It appears that, under Samoan conditions, cattle must be removed from mahogany plantations one month before the beginning of the wet season and kept away for several months. The silvipastoral system was also tried successfully in mixed mahogany, *Toona ciliata*, *Albizia* (=*Paraserianthes*) *falcataria* and *Securinega samoana* plantations, which produced 200 grazing days ha^{-1} yr^{-1} (at 400 stems ha^{-1}).

A trial silvipastoral system has been established in the western Amazon region of Brazil (Matos *et al.* in press). The mahogany component was set out at a spacing of 20 x 6 m. Along the rows, mahogany trees were mixed with *Schizolobium amazonicum* at 2 m intervals. Rows of *Inga edulis* were planted 2 m either side of the mixed mahogany rows, with a 1 m spacing along the rows (see Plate 7.3). Fodder species (*Desmodium ovalifolium* and *Brachiaria humidicola*) were planted in the 18 m wide strip between the three rows of trees species. Despite the fact that 73-81% of mahogany trees were damaged after 30 months, it is claimed that the rows of *I. edulis* acted as a physical barrier which reduced shoot borer attack (Matos *et al.* in press). It is not known if a pure mahogany control plot had been established to validate this claim. However, regardless of incidence of attack, it is possible that the presence of lateral shade has encouraged the mahogany to recover its form (see Section 11.1.1.3). The authors suggest that the system could be used on abandoned and degraded lands in the region.

Box 7.2. Mahogany mixtures in the Solomon Islands

Mahogany planted in lines through natural forest in the Solomon Islands has suffered severe damage owing to the prolific growth of a climber (*Merremia*), which invades from the inter-row bush (Anon. 1988a). This, coupled with the presence of the shoot borer, led to interest in planting mahogany on cleared sites with a nurse crop. A number of trials were therefore set up.

Approximately 17 ha of mahogany and *Terminalia calamansanai* were established at an overall spacing of 5 x 3 m (alternate rows), with the intention of removing the *T. calamansanai* at some (unspecifed) stage to release the mahogany (Anon. 1987b, 1988a). However, the rapid height growth and large spreading crowns of *T. calamansanai* created too much shade for the light-demanding mahogany. Given the size difference between the two species, harvesting and extraction of *T. calamansanai* would have caused excessive damage to remaining mahogany trees and the trial was therefore abandoned. Other fast-growing timber species (*Albizia* (=*Paraserianthes*) *falcataria*, *Anthocephalus chinensis*) were rejected for similar reasons.

In addition, 6 ha of mahogany and *Leucaena leucocephala* were established, again at an overall spacing of 5 x 3 m (alternate rows) (Anon. 1988a). From the point of view of the mahogany, the trial was successful: after 9 years, mahogany trees had 'above average' form. However, *L. leucocephala* was not considered to be an ideal species to mix with mahogany. Its susceptibility to the aggressive fungus *Phellinus noxius* and the recent arrival of the *Leucaena* psyllid (*Heteropsylla cubana*) to the Solomon Islands makes it a risky species to plant. *L. leucocephala* has no local value apart from use as firewood, which is not in short supply.

Securinega flexuosa, an indigenous species to the Solomon Islands, appeared to offer more potential (Anon. 1988a). The species is straightforward to manage in the nursery, easily established and quickly captures the site. It grows fast, has a dense, compact crown and a low stature which does not exceed 20 m at maturity. Mean stem diameter reaches 11 cm after 5 years (at 1200 stems ha^{-1}), with a dominant height of 13.6 m and excellent form. In a mixture, researchers believed that *S. flexuosa* would initially grow ahead of mahogany, encouraging the vertical growth of the latter. As *S. flexuosa* approached its maximum height, mahogany would be able to catch up with and overtop it. At this point *S. flexuosa* would be harvested, leaving a pure stand of mahogany. There would be problems associated with harvesting and extracting through the mahogany stand, but the small dimensions of *S. flexuosa* would make it an easy species to cut out. In addition, its tolerance to shade gives some flexibility over the timing of harvest, possibly permitting a delay until the mahogany has matured (Chaplin 1993). The timber is extremely durable and widely used for house posts in the Solomons (Whitmore 1984).

The design of a mahogany and *S. flexuosa* experiment is described by Anon. (1987b). A plot of *S. flexuosa* was established at 4 x 3 m in 1988. In 1990, alternate trees in alternate rows were cut out and the gaps were planted with mahogany. The spacing of the mahogany in the mixed stand was therefore 6 x 8 m. Adjacent to this plot, a control plot was established, containing pure, open-planted mahogany also at 6 x 8 m. The plots were assessed in 1990, 1991 and 1993, but the results have yet to be written up. Recent observations revealed that mahogany in the mixed plot had an average clear (merchantable) stem of 7-8 m, much higher than that of the pure control plot (Iputu pers. comm.). The raw data indicate that the mahogany was attacked by the shoot borer in both plots, but recovered its form in the mixed stand to a greater extent than in the pure stand.

When the experiment was designed, the intention was to thin out *S. flexuosa* gradually over years 3-10 and to thin the mahogany to its final stocking of 150 stems ha^{-1} at year 15. Unfortunately, the research branch of the Solomon Islands' Forest Division has insufficient resources to take further measurements or carry out further treatments and the plot is reported to be overgrown with creepers (Speed pers. comm.). Nevertheless, the experiment offers conclusive evidence that the form of mahogany can be improved by planting in mixtures. The careful selection of *S. flexuosa* has permitted the design of what is potentially an operationally feasible silvicultural system, although the mixture may be too management-intensive for large-scale plantation (Anon. 1988a).

Plate 7.3. Mahogany in a trial agroforestry system with *Schizolobium amazonicum* and *Inga edulis*. Western Amazon, Brazil. *Photo: R. Leakey*.

7.3 Planting in natural forest

Various techniques have been used to establish mahogany in natural forest. Weaver (1987) refers to these collectively as enrichment planting. 'Enrichment' is a suitable term to use when the desired species is planted at low densities or in discrete lines or parcels, with the long-term aim of maintaining a mixture of planted and natural forest species. However, the term is misleading and perhaps inappropriate for planted species which will dominate and eventually replace the natural forest. The term 'conversion' is therefore used here. Most examples of establishment of mahogany in natural forest fall into the conversion category. Two very different techniques have been used, which may be termed conversion line planting and conversion underplanting. There are relatively few documented examples of enrichment planting, as defined above.

7.3.1 Conversion line planting

The objective of conversion line planting is to convert degraded natural forest into valuable plantation at maturity, when the planted trees will constitute a closed canopy

(Weaver 1987) (see Plates 7.4 and 7.5). The technique produces less environmental damage than clearing forest for plantation establishment, but any benefits for biodiversity are likely to be short-term, since natural forest is gradually replaced by the plantation species. Conversion line planting can potentially cut the cost of a final crop to one third of what would be incurred by close planting on cleared sites (Dawkins 1965). However, experience in Puerto Rico indicates that this estimate of saving may be optimistic. Although line planting costs a little less than planting on cleared sites, plants on the latter are easier to maintain and give higher growth rates (Wadsworth pers. comm.).

Conversion line planting is only a suitable technique for certain species on certain sites. Dawkins (1965) studied experiences in francophone Africa (from Foury 1956) in order to produce 5 necessary conditions for successful conversion line planting (see Table 7.7). From the evidence presented in other sections of this review, it appears that mahogany is a suitable species for line planting if accompanied by appropriate silvicultural techniques. The main constraint to the use of mahogany for conversion line planting appears to be its lower than desirable height growth rate, particularly if this is further reduced by shoot borer attack. However, slower growth rates may be acceptable, particularly for a high value species such as mahogany, if forest managers are willing to pay for weeding and cleaning over a longer period.

Table 7.7. Meeting Dawkins' (1965) conditions for successful conversion line planting with mahogany

Conditions	Can the conditions be met?
1. The species to be planted must be self-pruning, straight and fast growing (>1.5 m yr^{-1}) light demanders	Mahogany has been known to achieve these height growth rates on good sites, where the effect of the shoot borer is not too debilitating. However, growth rates are typically 1 m yr^{-1} (see Section 9.2.1)
2. There must be no tree canopy; only clear-felled, clear-poisoned or low farm-bush conditions are suitable	Suitable canopy conditions are found in logged or secondary forests, although additional silvicultural treatment may be required to remove residual trees (see Section 7.3.1.1)
3. There must be little or no sale for thinnings in the area concerned	Although there may be local markets for mahogany thinnings, demand is usually low and most forest managers are only interested in growing large diameter mahogany for saw timber or veneer (see Section 8.5)
4. The regrowth or matrix between the lines must be non-inflammable	Rainfall is sufficiently high across most of the planted range of mahogany to make forest fire an unlikely event (see Section 6.1). However, sites with a pronounced dry season or near densely populated areas may be unsuitable
5. Browsing animals must be absent or of negligible effect on planted trees	Mahogany is unpalatable to most herbivores and browsing has not been a major problem (see Section 12.1.3)

The main advantage of growing mahogany in lines is widely perceived to be the protection afforded to the plants from the shoot borer, although some authors (e.g. Ikeda *et al.* in press, Matsumoto pers. comm.) strongly disagree (see Section 11.1.2.3). The wide spacing of lines may also limit the spread of root diseases, such as *Phellinus noxius* (Leslie 1994).

The following sections describe the key steps in conversion line planting and the various problems that have been encountered.

Plate 7.4. A young mahogany line planting. Note the natural vegetation between lines providing lateral shade. Luquillo Forest, Puerto Rico. *Photo: R. Leakey.*

Plate 7.5. A mature mahogany line planting. Note that mahogany crowns from adjacent planting lines have closed canopy. Galoa plantation, Fiji.

7.3.1.1 Removing the overstorey

The density of overstorey or canopy in logged or secondary forest is extremely variable, depending on the intensity of logging and/or the period of recovery. The less the canopy has been disturbed, the greater the work required to convert the forest to plantation. However, costs of overstorey removal may be offset by the sale of any remaining timber trees in the stand. In intensively logged forest, conversion is easier. Extensive mahogany plantations in Galoa, Fiji, were established on land that had been logged twice, the second time shortly before it was leased to the Forest Department (Kubuabola pers. comm.).

The overstorey may be removed by felling, girdling or poisoning (usually poison girdling). Poisoning has been the most commonly used technique for large-scale removal of the overstorey. In Fiji all non-commercial trees with a dbh over 30 cm (more recent reports suggest this figure may now be as low as 10 cm) are poisoned with a 20% solution of arsenic pentoxide in a frill girdle (Busby 1967, Oliver 1992). Some timber trees, such as *Endospermum* sp., *Agathis* sp. and *Dacrydium* sp., will be retained if they are of good form (Evo pers. comm.). Poisoning used to be carried out 3-6 months after the mahogany was planted (Busby 1967), but in some cases the poison was found to kill the mahogany trees, so poisoning is now carried out beforehand. For environmental reasons, arsenic pentoxide will soon be phased out by the Forest Department and other chemicals (Glyphosate, Butoxene, Ascot and Tordon) for tree poisoning are being tested (SRD 1995). A 10% solution of Tordon 50-D has been recommended, but trials are ongoing.

Researchers in Puerto Rico also studied the effects of poison girdling of natural forest species (Tropical Forest Research Centre 1959). Incised bark sprayed with 20% 2,4,5-T in water gave the best results over a 2 month period (16% dead), whereas basal spray with 5% 2,4,5-T in oil gave best results after a year (73% trees dead). Recent environmental concerns have led to the banning of lead arsenate and 2,4-D and the Forest Service is limited to the use of Glyphosate or Round-up (Francis pers. comm.). In Puerto Rico, these chemicals have not proved to be very effective, but results vary between countries. In Kolombangara, Solomon Islands, the overstorey (often consisting of *Albizia* sp.) has been effectively poisoned using Glyphosate or 2,4-D (Isles pers. comm.). A girdle is made and the poison squirted onto the stem on four sides of the tree. Poisoning is usually carried out before planting, but in some areas planting and poisoning are carried out simultaneously.

A longer-term poison girdling study in mixed mahogany, *Cedrela odorata* and *Cedrelinga catenaeformis* line plantings was carried out in Peru with interesting results (Maruyama and Carrera 1989). A total of 242 competing natural forest trees along nearly 7.4 km of line plantings were girdled with a machete (15-20 cm bands at breast height). Glyphosate solution was applied to the girdles in varying concentrations (33%, 25%, 17% and 0% or control). Mortality was measured after 2, 7 and 26 months. Of the 33 families, 20 had mortality rates of more than 80% by the end of the experiment, but the results (see Table 7.8) indicate that the poison takes a period of at least 7 months and possibly up to 2 years to work effectively. Although the most concentrated Glyphosate solution caused greatest mortality, the difference between treatments (excluding the control) was not significant. Differences in mortality did not vary significantly with dbh class (range 14-98 cm).

Table 7.8. Response of girdled trees to Glyphosate treatments (Maruyama and Carrera 1989)

Period (months)	Mortality (%) after different Glyphosate treatments			
	33% sol'n	25% sol'n	17% sol'n	Control
2	0.0	3.3	0.0	0.0
7	17.6	18.9	11.5	0.0
26	86.8	77.8	71.8	50.0

If chemical methods cannot be used, ring girdling is thought to be a more applicable technique than felling for large-scale application (SRD 1995). Francis (pers. comm.) advises cutting the girdle to a depth of 2.5 cm and a width of 10 cm (i.e. so that all cambium is removed) to have a high chance of killing the tree. However, girdles were found to kill only 42% of trees over a one year period (Tropical Forest Research Centre 1959). An experiment in Fiji produced no significant difference after 18 months between 4 different physical girdling techniques: notching (a ring cut 2-5 cm into the wood); frilling (a single line of overlapping axe cuts); double frilling (a double line of overlapping axe cuts); and peeling (stripping off a 20 cm width of bark) (SRD 1994). An average of 60% of treated trees were dead after a year. However, the effect of girdling depends very much on the size of the tree and the species (SRD 1995). Regirdling is recommended to improve results.

Felling of overstorey trees is time-consuming and expensive, although unlike the other techniques does give immediate and definitive results. Bauer (1987b) describes the cutting of overstorey trees in Puerto Rico with a chainsaw. In parts of the Solomon Islands, residual trees with a diameter of more than 20 cm are felled 6 months before planting (Oliver 1992, Chaplin 1993). Fruit, nut and *Ficus* trees with cultural significance are retained. Certain tree species in Western Samoa are retained for similar reasons (Oliver 1992).

7.3.1.2 Cutting lines

Countries with a long history of line planting with mahogany (e.g. Puerto Rico and Fiji) have traditionally orientated lines in an east-west direction, in order to maximise the amount of sunlight received by the young mahogany plants (Bauer 1987a, Oliver 1992). Planting lines run at right angles to the long sides of plantation boundaries in Western Samoa, but an east-west direction is preferred (Oliver 1992). However, if all overstorey trees are removed, an east-west orientation may not be important. Forest managers may in fact be willing to accept lower growth rates if other orientations give important advantages (see Table 7.9).

Table 7.9. Advantages of different enrichment line orientations (after Soubieux 1983)

Orientation of enrichment lines	Advantages
East-west	Allows plants to receive maximum light
Perpendicular to direction of dominant winds	Reduces likelihood of windthrow or windsnap
Parallel to contours	Reduces soil erosion
Perpendicular to direction of existing forest roads	Assists maintenance and extraction of thinnings

There may be a significant reduction in cost of forestry operations by orientating planting lines in relation to forest roads. This technique has been used in Fiji: lines have been cut in a herring-bone pattern towards the nearest extraction route (Oliver 1992) and, in the Galoa plantations, lines run along the direction of slope to meet roads running along ridge tops and/or valley bottoms (Evo pers. comm.). There is, however, an associated risk of soil erosion on the steeper slopes.

The purpose of line cutting is to allow overhead sunlight to penetrate, but to preserve side shade to prevent branching (Bauer 1987a). Soubieux (1983) argues that the important feature of the cut line is not its width at ground level but at canopy level. The higher the inter-line vegetation, the wider the line must be. Oliver (1992) describes opening of the canopy to produce a V-shaped cross-section, which is consistent with Soubieux' argument. In practice, the shape of tree crowns makes this kind of opening impossible without extensive pruning (Francis pers. comm.). Chaplin (1993) describes the opening of U-shaped lines in the Solomon Islands, which is probably a more accurate description of what can actually be achieved. Experience in Fiji indicates that lines with a wide opening at canopy level may not necessarily be advantageous and tall inter-line vegetation may actually encourage more rapid vertical growth (Singh pers. comm.).

Dawkins (1965) recommends a planting line width of about 2 m, with easy access along one side of the planted trees. In practice, the cleared width of planting lines is usually 2-3 m (see Table 7.10) although it is not always clear whether a similar opening is made at the level of the residual forest canopy. In Fiji, the lines are usually a little narrower and less distinct at canopy level (Evo pers. comm.). In contrast, on Kolombangara in the Solomon Islands some line plantings have so little vegetation between them that they almost resemble cleared sites (Isles pers. comm.). In some cases, this has been due to suppression of the understorey by a canopy of *Albizia* sp. (subsequently poisoned) and in others due to perhaps over-thorough cleaning by workers. Bauer (1987b) describes the removal of slash to the secondary forest between lines. However, the woody material may serve as a breeding ground for ambrosia beetles and pose a fire hazard in dry climates.

7.3.1.3 Planting density

Lines are typically spaced 9-11 m apart (see Table 7.10). This spacing broadly matches the final crop crown diameter (see Section 8.5.1) and indicates that all stands in the examples listed have the potential to become pure mahogany plantations, given suitable management. According to Dawkins (1965), plants spaced along the lines should give at least one-in-three selection of final-crop trees, sometimes one-in-four. Optimal final stocking for a light-demanding species such as mahogany has been predicted to be 100-150 stems ha^{-1} (see Section 8.5.1), although the figure may vary depending on preferred tree size at the end of the rotation. Observed stocking densities of line conversion plantings, typically 230-500 stems ha^{-1}, tend to be a little lower than Dawkins' recommendation. However, thinning operations are not always prescribed and a low stocking will be preferable if trees are to be left to self-thin.

Planting should be carried out soon after line cutting, which should follow overstorey poisoning (Dawkins 1965). Failure to time operations precisely may necessitate additional weeding and cleaning.

Table 7.10. Examples of mahogany conversion line planting from around the world

Density (stems ha^{-1})	Spacing between lines (m)	Spacing along lines (m)	Width of cleared line (m)	Pre-existing forest type	Notes on spacing between / along lines	Country	Source
231	9.4	4.6	nda	Logged natural and secondary forest	Wide spacing was adopted to avoid the necessity of thinning. Bush growing between lines was left to limit growth of mahogany branches	Honduras	Chable (1967)
234	9.0	1.0, 8.5, etc.	2.0	Logged natural forest	The spacing was considered unsuitable; a regular 4.5 m spacing along rows may have been better	Malaysia (Sabah)	Nasi & Monteuuis (1993)
278	9.0	4.0	2.0	Logged natural forest and failed plantations	Wide spacing has been adopted to avoid the necessity of thinning which was found to trigger ambrosia beetle attack. Spacing between rows is thought to be sufficient to prevent the spread of root fungus	Fiji	Oliver 1992, Evo (pers. comm.), Singh (pers. comm.)
333	10.0	3.0	nda	Logged natural forest	nda	Sri Lanka	Forest Department archives
333-500	10.0	2.0-3.0	3.0	Logged natural forest	The spacing is too wide, causing *Merremia* infestation. High weeding costs are incurred	Solomon Islands	Chaplin (1993), Iputu (pers. comm.), Isles (pers. comm.)
363-400	10.0 or 11.0	2.5	2.0 or 3.0	Secondary forest	The spacing is thought to be most suitable for secondary forest in Puerto Rico	Puerto Rico	Weaver & Bauer (1986), Bauer (1987b)
500	10.0	2.0	2.0	Logged natural forest	Even spacing is difficult to achieve because of the rockiness of planting sites. Baselines, spaced every 200 m, are planted with a different species (usually *Tectona grandis* or *Flueggea flexuosa*) to define compartment boundaries	Western Samoa	Oliver (1992)
662	13.7, 1.5, 1.5, etc.	1.5	4.6	Secondary forest	Unorthodox spacing between lines was based on a method originally designed by Brandis for establishment of teak in India. At least two mahogany plantations (Banhedawakka and Panirendawa) have been established in this way	Sri Lanka	Tisseverasinghe & Satchithananthan (1957), Sandom (pers. comm.)

nda = no details available

7.3.1.4 Problems encountered with conversion line planting

Verissimo *et al.* (1995) report the use of line planting by mahogany logging companies in southern Para, Brazil, to replace the felled trees. Every 10 m a 2-3 m wide swathe was opened up and mahogany planted at 10 m intervals. By 1992 around 4000 ha had been planted in this way. However, growth has been slow (less than 0.5 m yr^{-1} in height). Other planting lines in Brazil have given very poor results, with mahogany hardly growing at all (Anon. 1961, Palmer pers. comm.). Trees have been planted in 'tunnels' through relatively undisturbed natural forest and may receive no direct light. Where lines are cleared to canopy level, the tops often become spanned by lianes.

Problems arise from the failure of forest managers to remove the natural forest canopy, one of the key conditions for successful line planting set out by Dawkins (1965). Under conditions of low light intensity, a light-demanding species such as mahogany has little chance of reaching the overstorey, even if the undergrowth around the plants is completely cleared. Line planting in relatively undisturbed natural forest will not, therefore, produce good results.

Cases of unsuccessful conversion line planting have also been reported in Guadeloupe (Soubieux 1983), Honduras (Montesinos *et al.* 1985, Cruz pers. comm.), Malaysia (Appanah pers. comm., Kotulai pers. comm.) and Puerto Rico (Francis pers. comm.). In general, the problem has been one of poor maintenance.

7.3.2 Conversion underplanting

In contrast to line planting, the objective of conversion underplanting is to retain a shelterwood of natural forest species while the mahogany plants are young. Some authors believe that these conditions lessen the risk of shoot borer attack (see Section 11.1.2.3). The shelterwood usually consists of trees of uniform height at a sufficiently low density to produce a light or broken canopy. Most of the understorey is treated to encourage the growth of the mahogany, which is stocked at higher densities than in line plantings. Shelterwood trees are usually removed within a few years of plantation establishment. For best results, the shelterwood needs to be opened up soon after mahogany height growth starts to accelerate (Holdridge and Marrero 1940).

A study carried out by Ramos and del Amo (1992) on the effect of a shelterwood on survival and growth of mahogany produced results which have implications for conversion underplanting. In Veracruz, Mexico, seedlings of mahogany and 2 other species were planted at 1 m intervals, in three 50 x 20 m plots. The canopy over each plot received a different treatment, described as:
1. Open (68% light transmission); all canopy layers including herbs and shrubs were clear-cut and removed);
2. Top canopy (37% light transmission); only trees below 12 cm dbh were removed leaving the uppermost canopy layer; and
3. Medium canopy (17% light transmission); only trees above 12 cm dbh were removed, leaving the medium tree canopy layer.

Fastest height and diameter growth was obtained in the open canopy treatments, but the differences between top and medium canopy treatments were insignificant. Survival was highest in the top canopy treatment and maintained at a constant rate for the next 80 months. Survival in the open fell over the first 1.5 years but then remained constant. In the medium canopy treatment, survival was initially high but declined steadily over the

80 month period finishing very low (18% survival rate). It is clear that very shady conditions are inappropriate for mahogany, but there is a trade-off between growth rates and survival under lighter shade. On the basis of these results, the use of the top canopy treatment to maximise survival and minimise competition from weeds during the first year was recommended. Subsequently, the canopy should gradually be opened by girdling or felling of shading trees to allow the mahogany to grow rapidly.

Conversion underplanting has been used with success in a number of countries. In Puerto Rico, Holdridge and Marrero (1940) claimed that 'where mahogany is planted under a shelterwood, best results in growth are obtained'. A spacing of 2.5 x 2.5 m was recommended. Weaver and Bauer (1986) note that a mahogany plantation was established in 1963 using the underplanting technique. The overstorey was poisoned gradually and seeds were sown at a 3 x 3 m spacing. Lamprecht (1989) reports the successful establishment of mahogany under a canopy of *Albizia falcataria*. Conversion underplanting worked well in Sri Lanka, although only when the overstorey was rapidly removed after planting (Forest Department archives, unpublished). The shade created by widely spaced dominants of the old natural forest overstorey was found to reduce vine and weed competition without seriously retarding growth rates in young seedlings in Honduras (Chable 1967). Many of the mahogany taungyas in Belize (see Box 7.1) were established in a manner resembling conversion underplanting.

The silvicultural treatments required for successful conversion underplanting are rather vaguely described in the literature. Only Soubieux (1983) gives a detailed account. Two variations of the underplanting technique were tried in Guadeloupe at different stages:

- *Pre-1972*. All small diameter trees in degraded natural forest were cut and mahogany established at high densities (1100-1600 stems ha^{-1}) under a dense shade of large trees. The shelterwood was subsequently eliminated by poisoning. This variation of the underplanting technique was not very successful. Low light availability resulted in slow growth of the mahogany and encouraged the development of branches and forked stems. Poisoning of large overstorey trees was difficult and not always successful.

- *Post-1972*. There was therefore a shift to planting under lighter shade. This involved the felling or poisoning of all trees with a diameter of more than 30 cm. At least 50% of forest cover was removed. In addition, the undergrowth was fully cleared. Mahogany plants were spaced at 5 x 2.5 m giving a lower density of 800 stems ha^{-1}. Another 800 stems ha^{-1} of valuable natural species between lines were located and cleaned. Remaining small diameter trees were cut later or poisoned over the next few years. This variation was much more successful. Young mahogany plants grew well and the maintenance of some canopy cover has reduced competition from herbaceous vegetation and the need for weeding.

Nelson Smith (1942) also describes an underplanting technique that involves complete removal of mature canopy trees. It is suggested that mahogany is established amongst a young, naturally regenerated nurse crop of slower-growing species. A number of suitable Belizean species were identified in this context, including *Vochysia koschnyia*, *Virola merondonis*, *Aspidospermum megalocarpum* and/or *Calophyllum brasiliense*. The young shelterwood would provide the mahogany with more lateral and less overhead shade than a high canopy, perhaps affording some protection from the shoot borer and encouraging apical growth. Mahogany would eventually overtake the shelterwood trees, having achieved a suitable length of clear bole. The method was not widely adopted because of the difficulty of removing the more vigorously growing pioneer species and isolating appropriate nurse species. However, conversion underplanting did take place in some regenerating natural forest in Belize. For example, Cree (1957) reports the successful establishment of 12 ha of mahogany by underplanting a natural stand of 10-year-old *Bolotia* spp. in Chiquibul.

The suitability of the different methods described will depend largely on pre-existing forest structure and type. Whichever method is adopted, the costs of underplanting are likely to be high because of the intensive silvicultural treatment required. Costs may be offset by partial exploitation of existing forest (Soubieux 1983). However, despite successful experimental underplanting of silviculturally treated rainforest with mahogany in Queensland, Australia, the technique was considered uneconomic (Keys and Nicholson 1982).

7.3.3 Enrichment planting

The objective of enrichment planting is to increase the proportion of timber trees in degraded or secondary natural forest, while maintaining a proportion of natural forest in the long term. Mahogany may be established by line planting or underplanting, but at lower densities (less than 100 stems ha^{-1}) than required for natural forest conversion. Enrichment line planting was carried out during the 1980s in blocks of natural forest in Sri Lanka, with an inter-line spacing of 20 m (Sandom pers. comm.). The lines were poorly maintained and most of the mahogany has failed. One of the main disadvantages of widely spaced enrichment lines is that they can be very difficult to relocate (Palmer pers. comm.).

Enrichment planting may also be carried out in discrete areas of natural forest, usually on disturbed ground. Mahogany has been planted on skid trails in the Sinharaja rainforest of Sri Lanka, with apparent success (de Zoysa *et al.* 1986). It appeared to grow in the disturbed skid trail conditions better than many local species. However, because of fears that the species would invade the only remaining block of natural rainforest in Sri Lanka, many of the young stems have been cut out. Planting along skid trails has also been carried out in Quintana Roo, Mexico, to supplement naturally growing mahogany stocks (Negreros in press). Reported survival of seedlings is poor, at best 37%. Hassan (pers. comm.) believes that the soil on skid trails is too compacted for seedlings to survive. The planting of skid trails and log-landings has been tried in many parts of Quintana Roo, but the seedlings rarely receive treatment after planting and may not survive well (Gutierres and Cabral pers. comm.). It is both difficult and costly to manage such extensive linear plantations in natural forest.

There has been interest in planting clearings with mahogany to enrich natural forests in Honduras (Cruz pers. comm.). In Ami Santa Maria del Carbón, experimental plantings were established in 27 clearings to monitor growth rates and shoot borer attack. The plantings appear to have been successful, with only a brief period of shoot borer attack shortly after establishment. Clearings provide conditions of light similar to those in which mahogany regenerates naturally (see Section 2.5) and seedlings may establish more quickly than those in enrichment lines. In relatively undisturbed natural forests where much of the overstorey is still intact, planting in clearings will entail considerably less disturbance to residual forest than the creation of effective enrichment lines.

7.4 Summary

- Direct seeding is an effective means of establishing mahogany plantations. Two or three seeds should be sown per mound at a depth of about 2 cm towards the end of the dry season. The technique is recommended on well-drained soils where seeds are less prone to rot before germination.
- Pure mahogany stands (at 2-3 m square spacing) may be established if there is a local market for small-diameter mahogany thinnings. Young trees grow rapidly and early canopy closure will reduce the need for long-term weeding.
- Mahogany has been successfully planted in agroforestry systems, typically with maize or bananas. Mahogany trees will grow rapidly under such conditions. The use of longer-term agroforestry systems may be possible if mahogany is planted at low densities or if shade-tolerant perennial crops, such as coffee or cocoa are used.
- Plantation mixtures provide mahogany trees with a combination of lateral shade and overhead light, in order to encourage strong vertical growth. Compatible species should have a compact crown, a low top height and a short rotation, so they can be thinned out once mahogany has achieved an unbranched bole of satisfactory length.
- In mixed plantations, the difficulty of felling the short rotation length species without damaging residual mahogany trees may add to the cost of silvicultural operations considerably. Careful silvicultural manipulation will be required to prevent one species from dominating the other.
- Line planting is suitable for the conversion of logged or secondary forest to mahogany plantation. If mahogany trees are to grow vigorously, the residual canopy trees must be removed, leaving only understorey vegetation between planting lines. Poison girdling is the most efficient means of removing canopy trees, but it may take a year or more for some herbicides (e.g. Glyphosate) to take effect. The environmental impacts of all herbicides need careful consideration.
- Planting lines should be cleared to a width of 2-3 m and overhanging trees removed. Vegetation between lines provides suitable conditions of lateral shade and overhead light. A spacing between lines of around 10 m is usual, if the mahogany is to close canopy at maturity. Spacing along the lines is typically 2-5 m, depending on the thinning regime proposed.
- The technique of underplanting has been used effectively to establish mahogany plantations in degraded natural forest. A light shelterwood of natural species should be retained, but the undergrowth cleared to enable mahogany to be planted at high densities (5 m between lines, 2-5 m along lines).
- A shelterwood may consist of a high shade of residual canopy trees, or a low shade of more densely stocked understorey trees. The shelterwood must be thinned soon after planting to avoid suppressing the young mahogany trees. The silvicultural costs of conversion underplanting are high and the mahogany may be damaged when the shelterwood is removed.
- Enrichment planting may be appropriate if a proportion of natural forest is to be retained. However, widely or unevenly spaced enrichment lines are difficult to relocate and expensive to maintain. The overstorey above enrichment lines must be completely removed: in natural forest where much of the overstorey is still intact, establishment of mahogany in clearings may be a more suitable technique.

Chapter Eight
Plantation Maintenance

8.1 Weeding

Mahogany is considered to be more tolerant of ground weeds than many other tree species established in plantations (Evans 1992). Even so, frequent weeding on a 2-4 monthly cycle has been found to produce significant improvements in height growth (Marcano 1963, Weaver and Bauer 1986). Measurement of an 11-year-old plantation in Mexico showed that the highest growth rates were found where there had been clear weeding and line weeding (Vera 1989). The lowest growth rates were observed where there had been spot weeding. However, there is a danger that ground weeding increases the incidence of shoot borer attack (see Section 11.1.3.1). For example, Geary *et al.* (1973) found that when sites were weeded, the shoot borer ruined tree form. For this reason, some prefer the use of spot weeding, despite the adverse affect on growth rates that this technique may have (e.g. Matsumoto pers. comm., Villafranco pers. comm.).

A factor which contributes to the ground weed problem in mahogany plantations is the narrow, monopodial crown of young trees (Geary *et al.* 1973). Depending on planting density, the young canopy may cast insufficient shade to suppress weed growth for several years. In Puerto Rico, difficulties with weed control were such that there were doubts as to whether mahogany was a suitable plantation species (Geary *et al.* 1973). Close spacing, which encourages early canopy closure, will reduce the amount of weeding required. Marie (1949) suggested that the best way to keep young mahogany plantations free of vines is to raise food crops between the rows (see Section 7.2.2.1). Where mahogany is underplanted or line planted in natural forest, maintenance of shaded conditions may also be an effective way of reducing the need for ground weeding. However, some residual natural forest types act as a reservoir for weeds and associated plantations may require more weeding than cleared sites.

Lines in natural forest are particularly susceptible to creepers, lianes and scrambling weeds, which damage or suppress young trees. Creepers such as *Lantana*, *Scleria*, *Mikania* and *Merremia* cause serious problems (Evans 1992). In Fijian plantations, creepers are considered the main threat to seedling survival (Kubuabola and Evo pers. comm.). *Merremia* is a severe problem in the Solomon Islands (Oliver 1992). In Sri Lanka, *Mikania scandens* was found to be common in young plantations and needed to be rigorously controlled where small planting stock (less than 30 cm tall) was used (Perera 1955). Herbaceous vines (*Ipomea* spp.) in Puerto Rico climb stems rapidly and develop such weight of foliage that they can bend 7 m tall mahogany trees (Holdridge and Marrero 1940). Vines grow particularly vigorously on wet, fertile sites, continually suppressing the growth of mahogany trees (Geary *et al.* 1973).

The weeding regimes prescribed for young plantations around the world therefore reflect the natural vegetation found on the planting site (see Table 8.1). In open conditions, low frequency ground weeding is sufficient. Where mahogany is planted in lines through natural forest, ground weeding must be supplemented by frequent creeper cutting. Most forest managers carry out regular creeper cutting over a 5 year period, though Lamb (1966) recommends that creeper cutting should be continued throughout the rotation on a 4-5 year cycle. A balance must be sought between the cost of additional ground weeding or creeper cutting and the profits from higher timber volumes at the end of the rotation. For example, it is recognised that creeper cutting in Fijian plantations beyond year 5 would increase growth rates; yet future profits are not sufficient to merit additional costs (Evo pers. comm.).

Table 8.1. Weeding regimes used in mahogany plantations

Natural vegetation	Yr1	Yr2	Yr3	Yr4	Yr5	Notes	Country	Source
Cleared	2	2	2	1	1	*nda*	Malaysia (Sarawak)	Chai & Kendawang (1984)
Imperata grassland	3	2	2	0	0	Weeding (hoeing) begins 3-4 months after planting and continues until trees overtop grass	Philippines	San Buenaventura (1958)
Shelterwood of natural forest spp. (small dbh)	2	2	1	0	0	*nda*	Guadeloupe	Soubieux (1983)
Shelterwood of natural forest spp.	4	2	2	2	2	Weeding continues until year 7	Mexico (Veracruz)	Ramos & del Amo (1992)
Residual natural forest between planting lines	5	4	3	2	1	Weeding begins 3 months after planting. All weeds along the line are cleared	Fiji	Evo (pers. comm.)
Residual natural forest between planting lines	2	2	2	2	0	*nda*	Puerto Rico	Bauer (1987b)
Residual natural forest between planting lines	4	4	nda	nda	nda	Weeding to remove climbers and grasses	Grenada	Frederick (1994)
Residual natural forest between planting lines	7	4	3	2	1	All weeds along the line are cleared	Western Samoa	Oliver (1992)
Residual natural forest between planting lines	12	12	6	4	2	All weeds along the line are cleared 1, 2 and 3 months after planting. Spot weeding continues until mahogany seedlings reach 1 m tall. Weeding of *Merremia* until year 5*	Solomon Islands	Oliver (1992)

nda = no details available
* This frequency of weeding is not necessary in all parts of the Solomon Islands, e.g. 4 weedings per year are sufficient on Kolombangara (Isles and Viuru pers. comm.)

Weed control in mahogany plantations tends to be physical, rather than chemical. Many herbicides are not selective enough to avoid killing the plantation crop along with the weeds (Evans 1992). However, Nazif (1992) claims that application of 4 litres of Roundup (diluted in 500 litres of water) per hectare had no impact on mahogany, and was an effective means of controlling weeds such as *Imperata cylindrica* and *Merremia umbellata*. Biological weed control has been tried with some success: leaf miners have been used to reduce the presence of *Lantana camara* (although they were also found to attack teak); and parasitic *Cusenta chinensis* plants have had some success in controlling *Mikania micrantha* in Sri Lanka and India (Evans 1992).

8.2 Cleaning

Cleaning operations are most important where mahogany has been planted in lines through natural forest or on forest sites which have been recently cleared. In such situations there tends to be large quantities of woody regrowth, particularly coppice, which can quickly invade and suppress the mahogany (Evans 1992). Regrowth which does not overtop mahogany trees is not always unwelcome competition; lateral shade may encourage recovery from shoot borer attack (see Section 11.1.3.1). However, regrowth which does manage to overtop the mahogany crowns will suppress growth and adversely affect health.

The main objective of cleaning is to provide full overhead light for mahogany crowns by cutting or poisoning competing treees (Dawkins 1965). Fast-growing 'weed' trees arising between the lines which also pose a threat can be targeted, though much will depend on the financial resources available. The number of treatments depends on the growth rates of mahogany (vertically) and of the natural forest canopy (vertically and laterally across the line) (Bauer 1987b). In Puerto Rico, at least one cleaning operation is required to maintain the opening at canopy level (Francis pers. comm.).

Cleaning of planting lines is commonly neglected, often with severe negative consequences for plantation productivity (Sandom 1978, Francis pers. comm.). However, where overstorey removal has been thorough and planting lines properly cleared at the time of establishment, cleaning may not be essential. In Fiji, for example, there is no cleaning of woody plants after the initial poisoning, due to financial constraints on the Department of Forestry (Evo pers. comm.). Inter-line vegetation sometimes grows over the gap, but survival and growth of mahogany is generally acceptable.

8.3 Fertilising

A fertilising regime should be designed to enhance growth rates by correcting soil nutrient deficiencies. Fertiliser trials with mahogany seedlings (Yao 1981, Leslie 1994) and recommendations for fertiliser application to mahogany plantations are reported (ERDB 1996, Rodriguez 1996). However, the lack of information on the nutrient status of soils to which the fertilisers were applied limits the usefulness of these findings. Until more research has been carried out, fertilising regimes prescribed for mahogany plantations must be based on local experience.

Fertiliser has not been widely used in mahogany plantations, partly because of the tolerance of mahogany to nutrient-poor sites (see Section 6.2) and partly because of the high cost of the practice. However, fertiliser application in newly established plantations may help to reduce the costs of weeding by enhancing mahogany growth rate and reducing the period of susceptibility to weeds. Trials are currently under way in Fiji (Galoa) to assess the economic viability of this option (Evo pers. comm.).

Normally, fertiliser is only applied to seedlings. It may be added at the pole stage to boost response to thinning, and is occasionally used to increase growth over the years leading up to final felling (Evans 1992). Where the next crop is dependent on natural regeneration, fertiliser may be used to encourage flower and fruit production (Matthews 1992). It is unusual to fertilise tree crops continuously over a whole rotation, although an interesting exception is reported by a mahogany tree farmer in the Philippines (Rodriguez 1996). Large quantities of nitrogen fertiliser are applied to densely stocked stands throughout the rotation (see Table 8.2), producing 40-60 cm dbh trees in 15-20 years. Stands are maintained at very high stocking densities throughout the rotation (800 stems ha^{-1} at final felling) and productivity is extremely high. However, the quantities of fertiliser used seem excessive for any tree species.

Table 8.2. Fertilisation regime for fast-growing plantations (Rodriguez 1996)

Year	Fertiliser (NPK)	Quantity (g tree^{-1})	No. of applications	Quantity (kg ha^{-1} yr^{-1})
1	16:20:0	50	1	125
	46:0:0	100	4	1000
2	46:0:0	100	3	750
3	46:0:0	100	3	750
4	46:0:0	100	2	500
5	46:0:0	100	2	500
6	46:0:0	100	2	500
7-20	Urea	200	1	160

8.4 Pruning

Shoot borer attack may produce forked stems or 'witch's brooms' after continuous damage to flushing shoots. In open stands, mahogany is susceptible to epicormic shoot development until bark thickens, usually by year 8-10 (Chable 1967). For a high value timber species such as mahogany, pruning must be regarded as an essential operation. Yet, despite the widespread occurrence of branchy stems on young trees, only a few examples of the use of pruning to improve form have been reported.

Pruning of mahogany plantations was carried out on a large scale in the United Fruit Company's plantations in Honduras (Chable 1967). The mahogany trees responded very favourably, with wounds healing cleanly and rapidly. Pruning took place regularly over a 7 year period after establishment and branches were removed while they were small. A live crown measuring 35-40% of total height was left after pruning. Work was carried out during the early part of the dry season.

Chable (1967) concluded that, with judicious pruning, acceptable mahogany stands can be formed even where shoot borer damage is severe. However, the cost of pruning is high, particularly in dense plantations. Plantations in Honduras were established at a 9.4 x 4.6 m for this reason: treating a plantation at a 2.0 x 2.0 m spacing would have cost almost three times as much over the first 7 years. On the other hand, densely stocked plantations may encourage vertical growth and thus reduce the need for pruning.

Pruning is an essential part of the maintenance of mahogany plantations owned by Kolombangara Forest Products Ltd. in the Solomon Islands (Isles pers. comm.) (see Plate 8.1). Pruning is scheduled at 12-18 months (to a height of 2.5 m), 24 months (to a height of 4.5 m) and 36-48 months (to a height 7 m). A straight stem of 7 m or more produces at least one merchantable log. Pruning is also widely used in Forest Department plantations in Martinique, where virtually all stems in pure plantations are attacked by the shoot borer (Chabod in press). The form of final crop trees is good but silvicultural costs are high.

Chaplin (1993) believes that mahogany self-prunes satisfactorily and there is some evidence to suggest that mahogany will recover its form after shoot borer attack if the vegetation around the mahogany is carefully manipulated (see Section 11.1.3).

Plate 8.1. Young mahogany trees in a stand regularly attacked by the shoot borer which have been formed by pruning. Note the remains of the overstorey. Kolombangara, Solomon Islands.

8.5 Thinning

The main purpose of thinning is to reduce the number of trees in a stand so that those remaining have more space for crown and root development (Evans 1992). It is generally assumed that thinning leads to an increase in the diameter increment of remaining trees. The positive effect of thinning on the growth rate of mahogany is well documented. In Puerto Rico, a pure 27-year-old mahogany stand with a basal area which had been maintained at 25-35 m^2 ha^{-1} for many years had an average diameter growth rate of 1 cm yr^{-1}. In comparison, mahogany in a similarly aged (though mixed) stand which had been kept well thinned, with a basal area consistently below 23 m^2 ha^{-1}, had a growth rate of 1.5 cm yr^{-1} (Tropical Forest Research Centre 1959). Weaver and Bauer (1986) found that the diameter and height of 7-year-old hybrid mahogany at a basal area of 0-10 m^2 ha^{-1} grew slower than at a basal area of 10-20 m^2 ha^{-1} and significantly slower than at a basal area of more than 20 m^2 ha^{-1}. Valera (1962) compared the growth of a 0.45 ha unthinned stand (448 stems ha^{-1}) with a 0.36 ha thinned stand (308 stems ha^{-1}) of 28-year-old mahogany over a 5 year period. The diameter increment for the thinned stand was 32% higher than that of the unthinned stand over the period of study.

Thinning regimes which maximise per hectare productivity throughout the rotation do not necessarily maximise plantation profitability. This is particularly true for hardwoods such as mahogany, for which there is little demand for thinnings. On the international market, only saw and peeler logs over 50 cm in diameter have any value (Isles pers. comm.). Few local markets exist for small diameter timber, although notable exceptions include: the Philippines, where small stems down to 5 cm are saleable for cement wood board (Malvar pers. comm.) and young trees of 25-30 cm in diameter are much in demand for furniture (Castillo pers. comm.); Sri Lanka, where stems and even branchwood down to 15 cm are used for furniture (Ratnayake pers. comm.); and Honduras, where 15 cm diameter thinnings are saleable (Wheatley pers. comm.). In other countries, thinning regimes that favour rapid growth of selected final crop trees may therefore be considered most appropriate. However, it is important to ensure that selected trees are of good form. In stands which are threatened by the shoot borer, a regime that maintains a degree of competition between trees throughout the rotation will reduce development of lateral branches and encourage apical dominance (see Plate 8.2). A regime with a high stocking density (e.g. one that maximises productivity) may therefore prove to be a suitable model.

No suitable thinning regimes for mahogany have been empirically determined. The only thinning trials that have been identified (for mixed mahogany plantations in Sri Lanka) were abandoned before any useful results were produced (Forest Department archives, unpublished). Instead, examples of thinning and thinning regimes cited in the literature are based on interpretation of plantation inventory data. The regimes have been categorised by the stand types to which they are applied, i.e. pure plantations, mixed plantations and conversion line plantings. In fact, towards the end of the rotation, when the mahogany component becomes dominant, line planted and mixed stands may be managed in a similar manner to pure mahogany stands.

8.5.1 Thinning pure mahogany plantations

For high density, pure plantations (spaced at 2 x 2 m to 3 x 3 m), it is recommended that the first thinning takes place soon after 5 years (Marie 1949, Bascopé *et al.* 1957,

Anon. 1959) and no later than 10 years from the date of establishment (Marrero 1950, Tropical Forest Experiment Station 1950, Bascopé et al. 1957). Thinning should be conducted with care, for if the trees are released too quickly they have a tendency to form lateral branches (Anon. 1959).

Plate 8.2. A densely stocked mahogany stand showing well-formed stems. Near Kuala Lumpur, Malaysia. *Photo: R. Leakey.*

Thinning regimes have been proposed for high density mahogany plantations in Martinique, Indonesia and Sri Lanka (see Appendix I for stocking figures and Appendix V for associated yield tables) (Anon. 1975, Sandom and Thayaparan 1995, Tillier 1995). Prescribed stocking densities under the regimes of Martinique and Indonesia are broadly similar throughout most of the rotation, but differ significantly from the stocking densities prescribed for Sri Lankan plantations (see Fig. 8.1a). In the latter case, the first half of the rotation is characterised by high stocking densities but, as the stand begins to mature, heavy thinning takes place in order to increase growth rates and to encourage the development of crowns, seed production and natural regeneration. Basal areas are maintained between 30 and 50 m^2 ha^{-1} under the regimes of Martinique and Indonesia (see Fig 8.1b), but are significantly lower under the Sri Lankan regime because of lower diameter growth rates. In all examples, thinning is carried out at 7-10 year intervals during a 50-60+ year rotation.

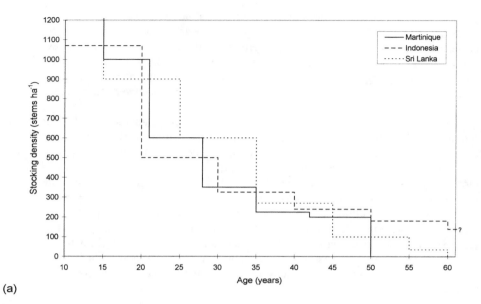

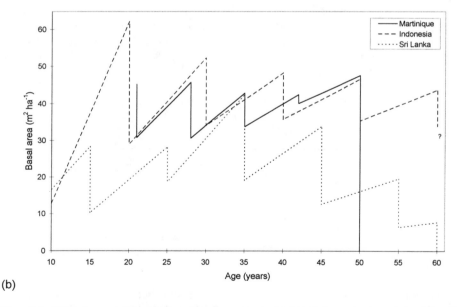

Fig. 8.1. Thinning regimes for mahogany stands in Martinique (using data from Tillier 1995), Indonesia (using data from Anon. 1975) and Sri Lanka (using data from Sandom and Thayaparan 1995): (a) stocking against age and (b) basal area against age.

Other authors favour basal areas considerably lower than those prescribed for Martinique and Indonesia. Lamb (1966) recommends that mahogany stand basal area should be kept below 23 m^2 ha^{-1} for rapid growth. Wright (1968) believes that stand basal area should be kept at around 22-24 m^2 ha^{-1}, although notes that a basal area of up to 29 m^2 ha^{-1} may be possible. These figures appear to support the thinning regime prescribed for Sri Lanka (see Fig. 8.1b). However, there is no evidence to show that the higher basal areas are associated with detrimental effects on plantation productivity. In fact, the thinning regime proposed for Martinique is the only one to have been put into effect over the course of a full rotation and the results have been very satisfactory (Tillier 1995).

The recommended stocking density for a mahogany plantation at the end of its rotation usually falls in the range of 100-150 stems ha^{-1}. Hummel (1925) predicted a final crop stocking of 100 stems ha^{-1} for plantations in Belize based on his experience with oak, which was considered to have similar light-demanding characteristics and crown morphology to mahogany. A final stocking of 140 stems ha^{-1} has been recommended in Indonesia (Anon. 1975) and 100 stems ha^{-1} in Sri Lanka (Sandom and Thayaparan 1995). There has been a tradition of light thinning in the French Antilles, producing slightly higher final stocking densities, e.g. 200 stems ha^{-1} in Martinique (see Fig. 8.1a). However, in order to increase growth rates and shorten rotation length, a final stocking density of 150 stems ha^{-1} has been been recommended by several authors (Anon 1959, Soubieux 1983, Chabod in press) and a heavier thinning regime may soon be put into practice. Tillier (1995) describes one possible scenario (see Table 8.3).

Mahogany tree farms in the Philippines have the highest reported final stocking density (Rodriguez 1996). The single thinning is carried out after 6-7 years, when plantation stocking density is reduced from 2500 to 1000 stems ha^{-1}. Natural mortality reduces stocking density over the next 10-15 years to 800 stems ha^{-1}, which represent the final crop. It should be noted that these plantations are located on soils with enhanced fertility through intensive fertiliser application (see Section 8.3).

Table 8.3. New thinning regime for mahogany stands in Martinique for average site class (adapted from Tillier 1995)

Age (years)	Stocking density (stems ha^{-1})	Type of thinning
5	1600	-
15	1000	Selective
21	500	Crown
28	250	Crown
35	150	Mixed
50	120	Regeneration

A number of authors have found a close relationship between stem diameter and crown diameter (see Table 8.4). Dawkins (1963) studied the stem diameter-crown diameter relationship for many tree species planted in the tropics. On the basis of an unspecified sample of mahogany, a linear relationship was derived. Wright (1968) took 20 sets of mean crown diameter and dbh measurements from 10 different even-aged mahogany stands in Fiji to determine the relationship between them. Stand age ranged from 3 to 28 years. The relationship that emerged between crown diameter and dbh was also linear and was said to correspond exactly with a relationship that had been defined for Puerto Rican mahogany trees. Samarasinghe *et al.* (1995) measured crown diameters of 60 mahogany trees of dbh from 5 to 60 cm and defined a curvilinear relationship between crown diameter and dbh, which has also been used to produce stocking figures. In this case, the relationship between dbh and crown width was highly significant, with a regression (r) value of 0.94. The equation described predicts a maximum crown width for a dbh of 180 cm, which is plausible but extrapolated well out of the range of collected data.

Table 8.4. Relationships between crown diameter and dbh

Equation (C = crown diameter)	Country	Source
C = 0.4 + 0.19.dbh	Unknown	Dawkins (1963)
C = 0.11 + 0.185.dbh	Fiji	Wright (1968)
C = 0.7466 + 0.2125.dbh - 0.0006.dbh^2	Sri Lanka	Samarasinghe *et al.* (1995)

By calculating how many crowns of a given size can be fitted into an area of 1 ha, it is possible to produce an acceptable estimation of associated stocking density. For this purpose, it is assumed that mahogany tree crowns are circular. Although reasonable for fast-growing young trees, this assumption may not be valid for older mahogany trees, but without further study it is impossible to say whether unevenly shaped crowns would significantly affect stocking predictions. It is also assumed that trees in plantations are set out at a square spacing, which is the normal arrangement for high density plantations (although triangular spacing would, in theory, allow a more efficient arrangement of circular crowns).

Stocking density is dependent not only on crown size, but also on crown coverage (which may be defined as the total area of every tree crown in a given compartment divided by the area of that compartment). Circular crowns may be positioned so that they are just touching (rather than overlapping with, or set apart from) neighbouring trees, giving a crown cover of 78.5%. From observations in Sri Lanka, this arrangement appears to be reasonable: mahogany crowns do not need space around them, but where branches interlock, injuries cause ill health and decay (Sandom pers. comm.). However, crowns may overlap without physical contact where neighbouring trees are of different height. In this situation, crown cover at a square spacing can exceed 78.5%. On the basis of measurement of trees in one sample plot in Fiji over a 5 year period from age 27 to 32 years, Wright (1968) demonstrated that a crown cover of 100% did not reduce annual increment in dbh and that mahogany could tolerate cover exceeding 100%.

The dbh-crown diameter equations that have been described (see Table 8.4) have been used to predict per hectare stocking densities and basal areas from dbh for hypothetical mahogany stands with both 78.5% and 100% crown coverage (see Appendix II for figures). The Sri Lankan equation (Samarasinghe *et al.* 1995) gives the lowest, and the Fijian equation (Wright 1968) the highest predicted stocking densities (see Fig. 8.2a). However, all stocking density curves converge to give a predicted stocking of 100-150 stems ha^{-1} when mean dbh reaches 50 cm (which, for most forest managers, represents the end of the rotation). The Sri Lankan equation also predicts the lowest basal areas until mean dbh reaches 50 cm (see Fig. 8.2b).

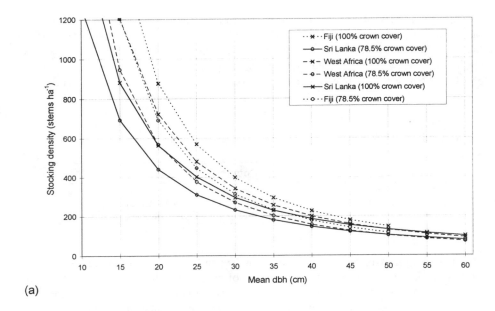

(a)

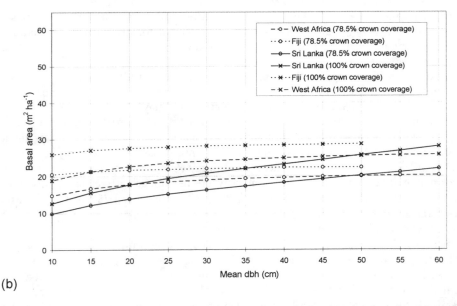

(b)

Fig. 8.2. Use of mahogany dbh-crown diameter relationships (after Dawkins 1963, Wright 1968 and Samarasinghe *et al.* 1995) to produce a guide to thinning: (a) stocking density against mean dbh and (b) basal area against mean dbh.

It is interesting to note that the crown diameter method produces stocking density and basal area predictions which are low in comparison with the prescribed thinning regimes of Martinique, Indonesia and Sri Lanka (see Figs. 8.3a and 8.3b). It appears that mahogany trees grown under the prescribed thinning regimes have either a smaller crown diameter:dbh ratio than the trees on which the dbh-crown diameter equations were based, or a crown coverage in excess of 100%. Given the light-demanding characteristics of mahogany, the former explanation is more likely, but more research into crown diameter:dbh ratios of densely stocked stands is required before any conclusions can be drawn.

The lack of agreement between prescribed thinning regimes and predicted stocking densities makes it very difficult to decide how mahogany stands should be managed. With evidence that some of the most productive stands have a basal area of around 40 m^2 ha^{-1} throughout much of their rotation, there is a strong case for recommending the thinning regime which has been successfully applied in Martinique. Yet it is probable that sites of low fertility, which are typical of many mahogany plantations, are unable to support such high basal areas. It could be also argued that less densely stocked stands will have a shorter rotation length, which may have important economic advantages. Taking the views of all authors into account, it is proposed here that a mean basal area of 25-30 m^2 ha^{-1} should be the target for forest managers, unless local evidence indicates otherwise. The stocking density predicted for Fiji (see Figs. 8.3a and 8.3b or Appendix II, Table II-B) should therefore serve as a reasonable guide to the design of an appropriate thinning regime (although it should be noted that stocking densities for trees under 20 cm dbh are unrealistic).

The type of thinning that is carried out is usually designed to maximise the size and quality of the final crop trees. Sandom and Thayaparan (1995) believe that selective thinning is most appropriate for mahogany stands, although the first thinning may be systematic. Where final crop trees are selected early (at year 14 in Martinique), stands are thinned from below to enhance the growth rate of dominant and sub-dominant trees (Chabod in press). Thinning from below may also be used to produce a more uniform spacing (Tropical Forest Research Centre 1957) and to remove any unhealthy trees which may harbour pests and diseases (Evans 1992). However, low thinning incurs considerable extra cost and may not significantly improve the growth of dominant or sub-dominant final crop trees. In Puerto Rico, it could only be justified in mahogany plantations where products could be disposed of at a profit (Tropical Forest Research Centre 1957).

Crown thinning has been used occasionally. The technique has the advantage of producing large, merchantable logs before the end of the rotation. Noltee (1926) advocated a crown thinning of early mahogany plantations in Indonesia. Some mahogany plantations in St. Lucia have been crown thinned to maximise growth of smaller, selected crop trees (Whitman no date). In one area (La Sorcière) thinning intensities (25-33% of basal area) were thought to be excessive. In the newly proposed regime for Martinique, two crown thinnings are recommended during the course of the rotation (see Table 8.3). However, there is a risk that crown thinning, if not carefully controlled, can lead to high-grading and loss of quality stems in the stand.

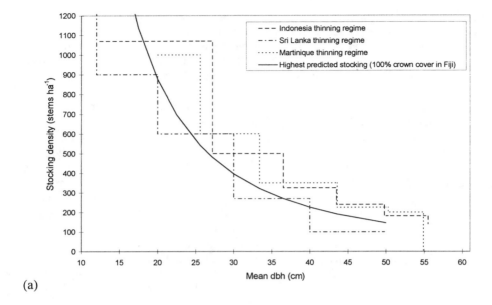

(a)

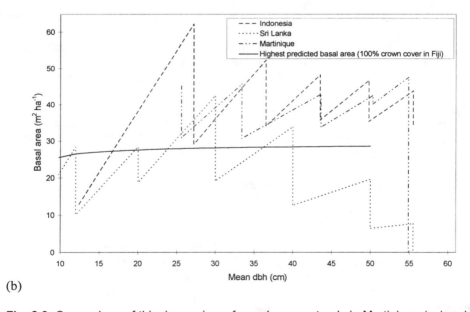

(b)

Fig. 8.3. Comparison of thinning regimes for mahogany stands in Martinique (using data from Tillier 1995), Indonesia, (using data from Anon. 1975) and Sri Lanka (using data from Sandom and Thayaparan 1995) with guiding curves derived from Fig. 8.2: (a) stocking density against mean dbh and (b) basal area against mean dbh.

8.5.2 Thinning mixed mahogany plantations

Although mixed mahogany stands are not unusual, examples of thinning practices have not been widely reported. Few mixed stands have been properly thinned.

There has been some experience of thinning mixed plantations in Sri Lanka. Mahogany and jak (*Artocarpus integrifolia*) plantations, established on a large scale during the 1920s and 1930s, were regularly thinned up until the 1960s. It was suggested at the time of planting that the mahogany might have to be cut back in the future, although it was anticipated that the shoot borer would check its growth (Anon. 1935). In fact, growth of mahogany was checked for only a short time. After two decades, jak was overtaken by the mahogany and started to die back (Tisseverasinghe and Satchithananthan 1957). Foresters took the view that jak should be favoured over mahogany, which meant cutting out good mahogany (crown thinning) in order to favour poor jak. The ratio of mahogany to jak, 3:1 at one stage, was reduced to less than 3:2 (Forest Department archives). However, thinning only favoured the remaining mahogany stems, which grew well under conditions of higher light intensity. The proportion of jak in the stands has declined ever since and it appears that the mixture of mahogany and jak is unsustainable. The example highlights some of the problems of designing thinning regimes for mixed stands.

In St. Lucia, selective thinning of mixed mahogany and blue mahoe (*Hibiscus elatus*) stands is being carried out to release suppressed mahogany (Bobb pers. comm.). Stands around 35 years old were still at the original spacing of about 2 x 2 m. The present objective is to maintain a ratio of mahogany to blue mahoe of about 6:1. Selective thinning is being carried out gradually to ensure that whip-like mahogany stems (though of good, unbranched form) are not damaged in strong winds. The mahogany is responding well to the treatment. Small diameter blue mahoe thinnings can be sold for fence posts but small diameter mahogany is unsuitable since the bark cannot easily be stripped off, preventing effective creosoting of the timber.

A thinning regime for a mixed plantation which may have general relevance is described by Soubieux (1983) (see Table 8.5). The plantations for which the regime was designed were originally established under a natural forest shelterwood. The remains of this were removed between 5 and 7 years, leaving a mixture of mahogany poles and similarly sized (carefully selected) natural regeneration. The objective of repeated thinnings is to raise gradually the proportion and quality of mahogany in the stand. Other species are retained throughout the rotation, though in ever decreasing proportion, to reduce possible incidence of shoot borer attack and other pests and diseases. This regime is particularly appropriate where there is no local market for mahogany thinnings and could serve as a model for other mixed plantations where mahogany is grown with a less valuable timber species or nurse crop.

8.5.3 Thinning mahogany line plantings

The cleaning of competing natural vegetation (see Section 8.2) is usually the main priority during the first half of the rotation of line planted mahogany stands, although an early thinning may be required if crown symmetry is to be maintained. Because of the low mahogany planting densities (usually less than 500 stems ha^{-1}), thinning cycles tend to be longer and thinning intensity lower than those in pure plantations. There is therefore little need to design complex thinning regimes for line planted stands. However, thinning operations are still vital to the health and productivity of the stand.

Table 8.5. Thinning regime for mixed mahogany plantations (Soubieux 1983)

Age (years)	Mahogany stocking (stems ha^{-1})	Natural spp. stocking (stems ha^{-1})	Total stocking (stems ha^{-1})	Objectives of thinning, other than spacing
10	800	800	1600	None
15	650	450	1100	Removal of unhealthy or branchy stems*
20	500	200	700	Removal of unhealthy or branchy stems*
30	380	80	460	Removal of trees with poor form
40	230	50	280	Removal of trees competing with final crop
50	105	35	140	Felling of crop trees leaving selected seed trees

* Non-commercial thinning

Bauer (1987a) recommends two thinnings (year 15 and year 30) for line planted (2.5 x 11 m) hybrid mahogany over the course of a 40-50 year rotation. In the faster-growing line plantings of the Solomon Islands, mahogany may close canopy as early as year 8 and in such cases a reduction in density from 333 stems ha^{-1} to 150-200 stems ha^{-1} is recommended at years 10-12 (Chaplin 1993). Line planted mahogany stands in Western Samoa will be thinned from 500 stems ha^{-1} to a final stocking of 100-150 stems ha^{-1} (Oliver 1992). In Fiji the original intention was to reduce stocking density to 100 stems ha^{-1} by the end of the rotation (Wright 1968), but the risk of ambrosia beetle attack was found to be high in thinned stands and thinning is no longer practised (Singh pers. comm.).

Unwanted natural species which have grown up between the planting lines must also be thinned. In some of the line planted stands of Indonesia, Malaysia, Sri Lanka, the Solomon Islands and Puerto Rico, where thinning has been neglected, many mahogany trees along the lines have become suppressed and the trees that attain dominance are often of poor form (Francis pers. comm., Kotulai pers. comm., Matsumoto pers. comm., Sandom pers. comm., Speed pers. comm.).

In general, the stocking densities recommended for pure plantations will be appropriate for maturing line planted mahogany stands (see Section 8.5.1).

8.6 Rotation length

Reported rotation lengths for mahogany stands are extremely variable (see Table 8.6), reflecting a range of management objectives. Where profitability is of primary concern, rotation lengths are generally as short as possible. This is particularly true of most privately owned plantations, in which rotation lengths of less than 40 years are the norm. Management objectives for state-owned plantations may be less profit-orientated and rotation lengths of up to 55 years have been reported. In the absence of available financial plans for mahogany plantations, it is impossible to consider the various economic variables, such as price of logs, maintenance costs and interest rates, which affect rotation length. Instead, the discussion here is limited to a brief consideration of growth rates.

Table 8.6. Suggested rotation lengths

Rotation length (yrs)	Plantation ownership	Country	Source
15	Private	Mexico (Tabasco)	Parraguirre (pers. comm.)
15-20	Private	Philippines	Rodriguez (1996)
25-30	State	Philippines	Castillo (pers. comm.)
30-35	Private and state	Solomon Islands	Chaplin (1993), Isles (pers. comm.)
30-50	State	Indonesia (Java)	Fattah (1992), Matsumoto (pers. comm.)
35-40	Private and state	Trinidad	Lamb (1955)
35-50	State	Fiji	Oliver (1992)
50	State	Martinique	Tillier (1995)
55	State	Belize	Ennion (1996)
55	State	Sri Lanka	Sandom & Thayaparan (1995)

Rotation lengths may be guided by the time taken for stands to reach their maximum mean annual increment (MMAI) in volume. Ennion (1996) predicts that MMAI will be reached at about 55 years for mahogany plantations in Belize. These plantations are found on sites of average quality, but have received minimal maintenance since establishment (self-thinning only). In Indonesia, Anon. (1975) predicts that mahogany plantations will reach their MMAI some time between 20 and 50 years, depending on site quality. This prediction is corroborated by Pandey (1983), who estimates that, on an average Indonesian site, MMAI will be reached at around 35 years. However, for plantations in which there is intensive fertiliser application, MMAI may be reached in 15-20 years (Rodriguez 1996). The drawbacks of using short rotations are high plantation maintenance costs and the poorer quality of timber produced (see Chapter 10).

There are a number of observations which may help to determine the *maximum* rotation length of plantation-grown mahogany. In Sri Lanka, trees over 75 cm in diameter were found to have a tendency to split on felling (Tissverasinghe and Satchithananthan 1957, Muttiah 1965). If large trees are present, careful delimbing (lowering branches to the ground on ropes) and felling (dropping trees onto beds of small branches) is recommended. In Martinique, observations from early plantations in both dry and humid parts of the island indicate that rotation lengths should not exceed 65 years if decline in timber quality is to be avoided (Marie 1949). The reason for the decline was not understood, but it may have been initiated by a hurricane which struck in 1928, 23 years after planting. Damage to crowns facilitates invasion of pests and diseases and it may therefore be wise to adopt shorter rotation lengths in hurricane-affected countries.

8.7 Summary

- On open planting sites, 2-3 weedings per year for 3 years may be sufficient for mahogany seedlings. In logged forest, creepers are often a serious problem and weeding of planting lines may be required 4-12 times during the first year, with a

- gradual reduction in intensity over a 5 year period. Spot weeding is the most suitable technique where the shoot borer is active.
- It is important to keep mahogany planting lines clear of competing understorey vegetation, in order to allow sufficient light to reach mahogany crowns. Occasional cleaning is beneficial (perhaps every 5 years) until mahogany crowns have reached canopy level.
- Although mahogany is tolerant of relatively infertile planting sites, application of an appropriate fertiliser will stimulate the growth of young mahogany trees on nutrient-poor sites. Long-term intensive fertiliser application may shorten rotation length substantially (perhaps to 15-20 years). However, the effect of different fertilisation regimes on growth rates and timber quality has not been properly investigated.
- Pruning is an effective means of improving the form of young trees, particularly those which have been attacked by the shoot borer, as long as it is carried out regularly over a period of 3-7 years after planting. Pruning is particularly important in low density plantations where there is little competition between crowns and thus less chance of natural recovery of vertical growth.
- The first thinning in pure mahogany stands at close (2-3 m) spacing should take place after 5-10 years. If trees have been attacked by the shoot borer, thinning should be delayed to encourage recovery of vertical growth and improved tree from.
- Reported thinning regimes and interpretation of dbh-crown diameter relationships indicate that suitable stand basal areas (over the course of the rotation) range from around 20 to 40 m^2 ha^{-1}. In general, maintenance of mahogany stands at a basal area of 25-30 m^2 ha^{-1} is recommended.
- Forest managers wishing to design new thinning regimes are referred to stocking density predictions for Fiji at 100% crown coverage (see Table II-B, Appendix II), which can be used as a guide. In order to obtain large crop trees, a final stocking of about 150 stems ha^{-1} is a suitable target.
- Thinning should be carried out on a 5-10 year cycle, depending on growth rates. The first thinning may be systematic, but with the emphasis on producing well-formed final crop trees, selective thinning is recommended over the rest of the rotation. If stands are stocked with a sufficient number of well-formed trees, it may be possible to take crown thinnings before the end of the rotation.
- Thinning regimes for mixed mahogany plantations need to be designed to balance the light requirements of all species involved. If there is no local market for mahogany thinnings, it will be preferable to target non-mahogany species for thinning early on in the rotation, which will help to release the light-demanding mahogany. It may be necessary to thin out the non-mahogany species before the end of the rotation, depending on their compatibility with mahogany.
- Thinning conversion line plantations is straightforward because of the low mahogany stocking densities (typically less than 500 stems ha^{-1}). A maximum of two thinnings is required; these operations should aim to remove both mahogany and unwanted natural species which have grown up between planting lines.
- Reported rotation lengths vary from 15 to 55 years. Large diameter trees (over 50 cm in dbh) may be produced in under 20 years with fertiliser application. There is evidence that a rotation length of 20-30 years is feasible on fertile sites. However, for plantations stocked at densities that are sufficient to encourage the development of acceptable tree form and timber quality, a rotation length of 30-50 years is anticipated, depending on site quality.

Chapter Nine
Growth and Yield

The ability to predict the growth and yield potential of mahogany plantations is of considerable importance for plantation planning, but few countries have enough reliable data on which to base such predictions. Forest managers are therefore faced with two options: to establish sample plots and initiate growth and yield studies; or to utilise growth and yield data from other mahogany stands in similar environmental conditions (Pandey 1983). The aim of this section is help those who are unable to establish sample plots, or who need to make growth and yield predictions before sample plot data become available.

Available growth and yield data are discussed and relationships between key variables described (see Section 9.1). All empirical growth functions that have been reported in the literature are also described (see Section 9.2 and 9.3). Unfortunately the quality and quantity of available data were inadequate for validation of most of the growth functions.

Abbreviations have been used in tables throughout this chapter (see Table 9.1 for details).

9.1 Available growth data

Raw growth data from mahogany plantations are not widely available. The research and, in some cases, commercial value of such data may be high and forestry institutions are understandably reluctant to divulge the results of many years of careful measurement. It was therefore necessary to collate processed data, in the form of pre-calculated stand parameters (mean dbh, stocking density, dominant height, etc.) from published literature. In-country 'grey' literature, which sometimes contains summaries of continuous measurements in permanent sample plots, was found to be a richer source of growth data than literature with a wider circulation. Inevitably, the quantity and quality of processed data are low. However, this problem is by no means unique to mahogany and reviews of other plantation species have revealed a similar paucity of published information (e.g. Somner and Dow 1978, Pandey 1983). The challenge is to make the best use of the data that are available.

Processed mahogany data have been classified according to the nature of the inventories from which they were derived. Two types of inventory have been recognised:

- *Single inventories* (one set of stand measurements from a given sample area). Stand parameters from such inventories have been reported from 21 countries around the world (see Appendix III). The sample areas include temporary sample plots, trial plots, research plots or even whole stands. In some cases, reported parameters represent the means of sample plots over a large plantation area.
- *Continuous inventories* (two or more sets of stand measurements from the same sample area). Stand parameters from such inventories have been reported from 10 countries around the world (see Appendix IV). The sample areas include permanent or semi-permanent sample plots (see Plate 9.1).

Both single and continuous inventory data had many limitations and it was necessary to make certain assumptions in order to use the available data to their full potential (see Section 9.1.1). Subsequently, it was possible to describe relationships between key variables (see Section 9.1.2).

Table 9.1. Abbreviations used in this chapter

Abbreviation	Definition (units)
For mahogany stands*	
D_m	mean dbh (cm)
D_g	dbh of the tree of mean basal area (cm)
D_d	dominant dbh (mean dbh of the 100 trees of largest dbh per hectare) (cm)
G	basal area (m^2 ha^{-1})
H_m	mean height (m)
H_d	dominant height (mean height of the 100 trees of largest dbh per hectare) (m)
A	age (years)
N	stocking (stems ha^{-1})
S	site index value
V	volume (no further specifications given) (m^3 ha^{-1})
V_x	volume to a top diameter of x cm (no further specifications given) (m^3 ha^{-1})
V_{xo}	volume to a top diameter of x cm overbark (m^3 ha^{-1})
V_{xu}	volume to a top diameter of x cm underbark (m^3 ha^{-1})
V_{MAI}	volume mean annual increment (no further specifications given) (m^3 ha^{-1} yr^{-1})
V_{CAI}	volume current annual increment (no further specifications given) (m^3 ha^{-1} yr^{-1})
For single mahogany trees	
D	dbh (cm)
H	height (precise definition given in associated table) (m)
V	volume (precise definition given in associated table) (m^3)
V_m	mean volume (m^3)

* Lower case abbreviations used for thinnings

9.1.1 Limitations and assumptions

The data (in the form of stand parameters) that are described vary considerably between sources (see Appendices III and IV). Age and mean dbh are usually reported. Mean heights are often reported for young stands, but mean and dominant heights of older stands are reported less often, presumably because of the difficulties involved in taking measurements. Site quality has only been described in terms of dominant height, which has been used to define site classes; because no independent assessment of site quality has been made, the classes reported do not provide additional information (and have therefore been omitted from Appendices III and IV). Mahogany stocking density and/or total stocking density is often reported, but the stocking density of any non-mahogany component in the stand is often ignored. Stand basal area, volume estimates and mean annual increment are rarely reported. In no cases have the number of stems, basal area, or volume removed by thinning been reported. Most measurements have been taken in mahogany stands under 40 years old.

Because many stand parameters are inter-dependent, the failure to record one parameter may make it difficult to define a clear relationship between others. For example, mean dbh is strongly influenced by stocking density (and therefore thinning practices) and site quality; any relationship between mean dbh and age which fails to take these variables into account will have little practical value (Kriek 1970). In addition, the lack of information on variables such as stand maintenance, seed origins, sampling procedures, mensurational method, limits of precision and accuracy, and volume concept further reduces the quality of the data. As a result, the comparability of data from different countries is greatly compromised.

In many cases, the stands from which the data were taken are known or suspected to contain a non-mahogany component. The lack of information on the stocking density and/or basal area of non-mahogany species makes it difficult to assess their impact on mahogany growth rates. Much depends on the quality of plantation management. With careful silvicultural manipulation, a significant proportion of non-mahogany trees may be retained in a stand without adversely affecting the growth of the mahogany. Other species may assist mahogany by encouraging vertical growth and recovery from shoot borer attack. Unfortunately, silvicultural details rarely accompany stand parameters. Where there is no evidence to the contrary, it has been assumed that the non-mahogany component of a stand has no significant effect on the growth of mahogany. If such a stand does in fact contain a substantial or dominant non-mahogany component, any relationships produced will tend to underestimate the potential mahogany growth rate.

The shortage of estimates of volume and mean annual increment (MAI) for mahogany stands makes it difficult to assess their potential productivity. In cases where stocking density and mean dbh have been reported, stand volume and thus MAI have been calculated in order to enhance the usefulness of data. Volume and MAI figures have been calculated using a single entry tree volume equation which is thought to be generally applicable (see notes in Appendix III for further details). In estimating MAI, it has been assumed that no thinnings were taken. This is probably a reasonable assumption: the stands are young (the great majority are less than 35 years old) and the volume of merchantable thinnings likely to be low; many of the data are taken from line planted stands, which yield few mahogany thinnings until late in the rotation.

Unfortunately, a significant proportion of the continuous inventory data were rejected for the following reasons:
- parameters were derived from stands which had been subjected to unorthodox silvicultural treatments;
- parameters were derived from stands which had been managed in a way that reduced the long-term productivity of mahogany;
- non-standard parameters were calculated.

Further details are given in notes accompanying the data (see Appendix IV). It should be stressed that although the data are of little general use they may have local value.

Plate 9.1. Permanent sample plot CR20 (plantation established in 1958). Columbia River Forest Reserve, Belize.

9.1.2 Graphical relationships

Despite the limitations described, many of the reported (and derived) stand parameters are thought to be of sufficient quality to permit the definition of relationships between them. All relationships are shown graphically and are limited to those which are (i) based on a sufficient number of data points and (ii) are most likely to be of interest to forest managers (see Figs. 9.1-9.5). To improve the quality of the relationships, data have been stratified by country or stocking density.

In the absence of any information on site quality, stratification by country was considered the only alternative. Replacement of the independent variable (age) with dominant height in the relationships described would have eliminated the need for site quality stratification, but would also have greatly reduced the amount of data available for analysis and was therefore ruled out. Although stratification by country is a crude technique, it should be remembered that the growth rates of mahogany are influenced by factors other than site, such as local management techniques and shoot borer populations; stratification by country is therefore likely to reduce the variation caused by these factors.

Stratification by stocking density was used to reduce variation in mean dbh and mean height caused by competition from neighbouring trees. Mean dbh and mean height are affected by the pattern of stocking density through time. However, in the majority of cases, no stand histories were available and the stocking density used for stratification was the density which was recorded at the time of measurement. Given the quantity of the data, further stratification was impossible.

The graphical relationships that have been produced (see Figs. 9.1-9.5) are self-explanatory. Even after stratification, the quality of the relationships is poor. Using simple mathematical functions, straight lines or curves have been fitted to the data (where appropriate) in order to emphasise the trend of the relationships. These functions have been included for general interest only and cannot be used with any confidence to predict growth and yield without validation, except in the countries to which they specifically apply.

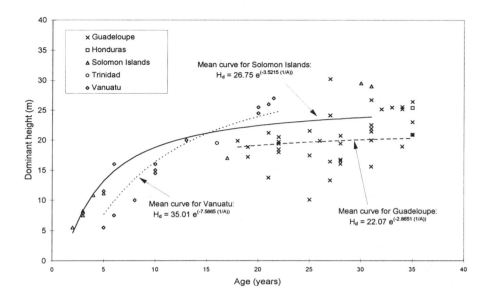

Fig. 9.1. Dominant height against age for different countries. Data from single inventories of sample plots in Guadeloupe (37), Vanuatu (13), Solomon Islands (9), Honduras (2) and Trinidad (1). See Appendix III for data and sources.

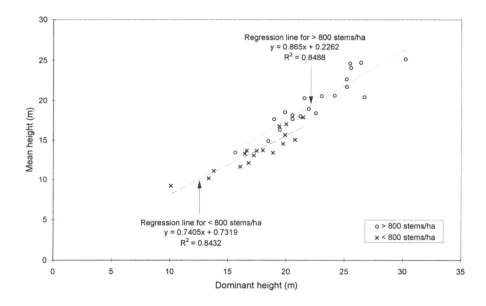

Fig. 9.2. Mean height against dominant height for different stocking classes. Data from single inventories of 37 sample plots in Guadeloupe (Soubieux 1983). See Appendix III for data and sources.

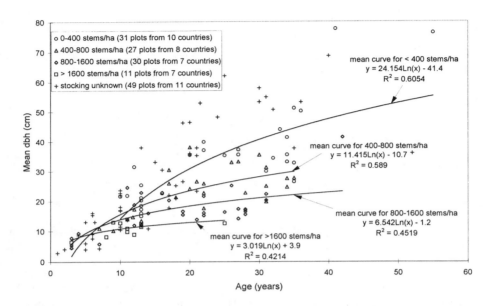

Fig. 9.3. Mean dbh against age for different stocking density classes. Data from single inventories of 148 sample plots in 19 countries. See Appendix III for data and sources.

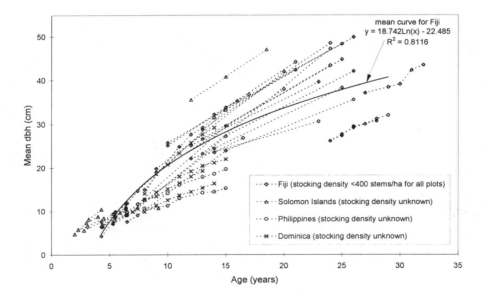

Fig. 9.4. Mean dbh against age for different countries. Data from continuous inventories of sample plots in Fiji (25), Solomon Islands (5), Philippines (4) and Dominica (2). See Appendix IV for data ranges and sources.

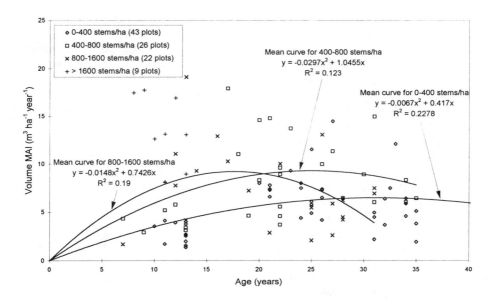

Fig. 9.5. Volume mean annual increment against age for different stocking density classes. Data from sample plots in Guadeloupe (37), Fiji (30), Puerto Rico (11), Belize (8), Nicaragua (6), Honduras (3), Mexico (3), Brazil (1), Panama (1) and Peru (1). See Appendix III and Appendix IV (Fiji only) for data and sources.

9.2 Predictions of growth and yield

The growth functions described in this section are derived from measurements of stands where mahogany is the dominant species, in terms of both crown position and stocking density. The functions are therefore only relevant to stands where mahogany is being managed as the principal crop species. Such stands may include planting lines if the overstorey has been fully removed and competing natural vegetation properly cleaned. In fact, many data and functions have been taken from well-managed line planted stands in Fiji. However, the functions are likely to overestimate growth rates if applied to mahogany in poorly managed line planted stands where there is more competition from non-mahogany trees species.

All of the relationships described in this section are derived by the authors to which they are attributed. Some of the relationships are simple hand-drawn curves with no associated empirical functions or statistical analysis. The usefulness of such curves is limited: predictions are difficult to read off and the relationships described cannot be incorporated into yield models. Relatively simple (third order polynomial) functions have therefore been fitted to the curves. The polynomial functions used are sufficiently flexible to match the predicted growth pattern very closely. No statistical measures of goodness of fit have been included. Such measures would be meaningless because functions have been fitted to predictions (derived from other functions or stand averages) rather than the original data. It must be remembered that the polynomial functions are valid only over the parameter range described. *Extrapolation outside this range is likely to give unrealistic or impossible values and should be avoided.*

Confidence limits are not given for any of the predictions reported in the literature and cannot be derived without the original data. The precision of the predictions is therefore unknown. However, it is possible to obtain a qualitative impression of precision by studying the number and distribution of sample plots from which data was taken (if the information is available). Somner and Dow (1978) used the term *'reliability'* and suggested a simple three-way classification:

- Reliability class 1: growth and yield data representative of large plantation areas (yield tables and large-scale inventories).
- Reliability class 2: growth and yield data from a varying number of sample plots, reliable for the stands of respective plots but not representative for large plantations.
- Reliability class 3: growth and yield data which are only unspecified estimates for sample plots or plantations of varying areas.

All growth functions described in this section are assigned a reliability class using available information on the distribution of sample plots, the number of sample plots and the number of remeasurements. Readers should be confident of the precision of functions that have been assigned a reliability class 1, but should treat functions assigned a reliability class 3 with considerable caution.

Although a function may be reliable, it may not be *valid* for predicting growth rates in other countries where the interaction of different climate, sites and provenance may produce a very different growth pattern. Validation of a growth function is possible if the function can be tested against independent data. Unfortunately, validation of all but two functions (see Section 9.2.1.3 and 9.2.2.2) was impossible using available data (see Section 9.1). Where validation was possible, it was accomplished by graphical analysis of residual errors (Alder 1980).

If forest managers intend to adopt any of the growth functions described, *it is strongly recommended that they are validated using any data that are locally available.* If no such data are available, it is advisable to match local conditions of climate and site as closely as possible to those in the plantations from which the growth curves were derived (see Chapter 6 for relevant information).

9.2.1 Height predictions

9.2.1.1 Mean height

Predictions of mean height growth are of most value in young plantations before canopy closure, when they may be used to assist with the planning of weeding and cleaning regimes. In addition, researchers and forest managers interested in growing mahogany in a mixture may wish to predict the mean height of young stands when trying to identify compatible tree species.

A number of functions describing the relationship of mean height with age have been described (see Table 9.2a). Because the functions are only valid for young plantations, and thus a short part of the overall rotation, the functions are based on linear or simple curvilinear relationships. Functions describing the relationship between mean height and mean dbh have also been proposed (see Table 9.2b).

The functions take no account of site quality, which limits their general applicability. The unpredictable effect of shoot borer attack on the growth of young mahogany trees also makes precise prediction of mean stand height extremely difficult. Persistent damage may produce trees with bushy form and slow height growth rates. Natural recovery of apical dominance may follow, but will depend on many factors (see Section 11.1.1). However, it would be unrealistic to expect a height growth function to include the shoot borer variable, because of the great variability in incidences of shoot borer attack.

Table 9.2a. Relationships between mean height and age for young mahogany

Growth function	Country	Region	Range	R	Source
$H_m = 10^{(1.6768-3.708A^{-1})}$	Fiji	Viti Levu	3-7 yr	2	JICA (1982)
$H_m = 0.271+1.1A$	Mexico	Tabasco	0-7 yr	3	Llera & Melandez (1989)
* $H_m = 0.01040A^3-0.0467A^2+1.165A$	Vanuatu	Many isles	1-15 yr	2	Neil (1986)

See Table 9.1 for abbreviations R = reliability class
* Function fitted to a hand-drawn curve

Table 9.2b. Relationships between mean height and dbh for mahogany under 10 yrs old

Growth function	Country	Region	Range	R	Source
$H_m = 1.2D$	Puerto Rico	Island wide	0-10 cm	2	Holdridge & Marrero (1940)
$H_m = 0.49+1.026D$	Puerto Rico	Luquillo	0-10 cm	3	Bauer (1987a)

See Table 9.1 for abbreviations R = reliability class

9.2.1.2 Dominant height and site index curves

The relationship between dominant height and age has been used by a number of authors to define a set of site index curves for mahogany plantations (see Table 9.3 and Figs. 9.6a-f). The purpose of the curves is to provide a means of assigning site index values or site quality classes to existing plantations on the basis of dominant height measurements. In most cases the curves follow the mean pattern of dominant height-age data points from all sites. Only Tillier (1995) stratified dominant height-age data by an independent assessment of site quality (altitude and rainfall) to define a set of polymorphic site index curves (see Fig. 9.6a). Even so, it is still not possible to predict the pattern of dominant height growth from a knowledge of specific site characteristics.

The functions described are sigmoidal or logarithmic, producing site index curves of a variety of shapes. It is likely that some of the relationships defined will be applicable to mahogany plantations in other countries. The spread of the curves can be adjusted by inserting different site index (S) values into the relevant equation (see Table 9.3). Sigmoidal funtions (see Figs. 9.6a,c,d,f) are thought to describe the pattern of natural growth most accurately, but if measurements from older stands are not available, a logarithmic curve (see Figs. 9.6b,e) may give a better fit to the data. However, the logarithmic functions which have been fitted to data from young stands are slow to reach an asymptotic value and therefore tend to overpredict growth rates for mature trees.

For comparison, readers are referred to Section 9.1.2 (Fig. 9.1) for dominant height-age data from other countries and associated functions.

Table 9.3. Relationships between dominant height and age for mahogany

Growth function for site index curves	Range of S values	Index age	Growth function for mean dominant height curve	Country	Region	Range	R	Source
*$H_d = 1.2088A - 0.0111A^2$ *$H_d = 0.9397A - 0.0076A^2$	best site worst site	-	not applicable	Belize	Columbia River	5-38 yr; 5-29 m	2	Ennion (1996)
†$H_d = S(10^{-(2.7373(A^{-1}-0.1))})$	10-25	10	†$H_d = 10^{(1.2304-2.7373(A^{-1}-0.1))}$	Fiji	Nukurua	5-25 yr	1	Wescom (1979)
$H_d = 2+(0.87S+3.17)(1-e^{-(0.048-0.88S^{-1})A^{1.30}})$	25-37	50	not applicable	Martinique	Pitons du Carbet	10-50 yr	1	Tillier (1995)
$H_d = S(40A^{-1})^{-0.76605}$	10-35	40	$H_d = 1.4906A^{0.76605}$	Philippines	Many islands	5-55 yr; 10-35 m	2	Revilla et al. (1976a)
$H_d = S(0.0908A^{0.6746755})$	15-35	35	$H_d = 2.5655A^{0.6746755}$	St. Lucia	Island-wide	10-32 yr; 9-32 m	2	Andrew (1994)
$H_d = S(2.0456e^{-(20.28386A^{-1.4})})$	8-20	10.9	$H_d = 26.843e^{-(20.28386A^{-1.4})}$	Sri Lanka	Intermediate zone	7-69 yr; 6-34 m	2	Mayhew (in press, a)

See Table 9.1 for abbreviations R = reliability class
* Function fitted to hand-drawn curves
† Pre-dominant height (height of the 20 trees of largest dbh per ha)

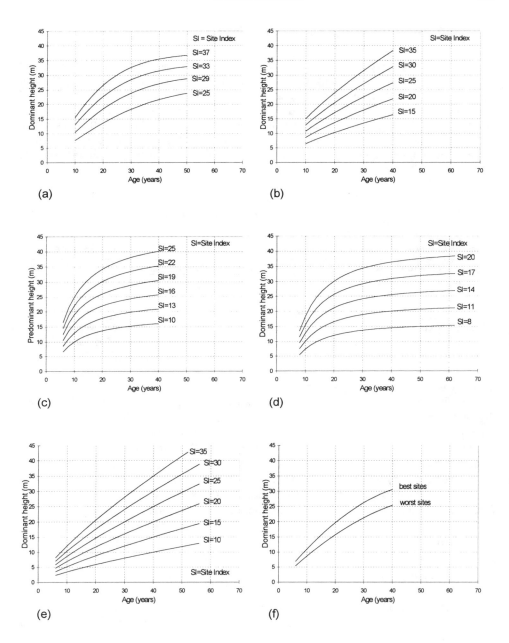

Fig. 9.6. Dominant height against age (see Table 9.3 for equations and other details) for: (a) Martinique, index age = 50 years (Tillier 1995); (b) St. Lucia index age = 35 years (Andrew 1994); (c) Fiji, index age = 10 years (Wescom 1979); (d) Sri Lanka, index age = 10.9 years (Mayhew in press, a); (e) Philippines, index age = 40 years (Revilla 1976a); (f) Belize, no index age given (Ennion 1996).

9.2.1.3 Relationship between mean and dominant height

Tillier (1995) suggests a relationship between mean and dominant height, which incorporates the stocking density variable (see Table 9.4). The intended application of this relationship is not described by the author. However, an estimate of the value of mean height is useful for the prediction of stand volume from single tree (double entry) volume tables. The use of dominant height (where there are more than 100 stems ha^{-1} in the stand) would give an overestimation of stand volume. Conversely, there may also be some value in being able to predict dominant from mean height to give an approximate idea of site quality.

Table 9.4. Relationship between mean height and dominant height for mahogany

Growth function	Country	Region	R	Source
$H_m = 0.846H_d + (692.7/N) - 0.0009N + 0.489$	Martinique	Pitons du Carbet	1	Tillier (1995)
See Table 9.1 for other abbreviations				R = reliability class

The range of values for which this linear relationship is valid are not described. For low values of dominant height, when stocking is also low, predicted mean height is greater than predicted dominant height: an impossible outcome in reality. The nature of the relationship is such that, as stocking density increases, the increase in predicted mean height with increase in dominant height tends to zero.

The function was validated using data from Guadeloupe. The plot of residuals (see Fig. 9.7) indicates that the function is biased, overpredicting mean heights by 1-2 m on average, with poorest results for mean heights of 15-20 m. Residual analysis indicates that the function needs adjustment to remove the bias. However, with so little validating data, it is difficult to say whether the function has any general applicability.

For comparison, readers are referred to Section 9.1.2 (Fig. 9.2) for a description of mean height-dominant height data from Guadeloupe and associated functions. These functions are simpler and may be more appropriate for prediction.

9.2.2 Diameter predictions

9.2.2.1 Mean diameter

If stocking density is known, mean dbh figures can be used to calculate basal area and volume for a stand of given age in order to give an idea of productivity. Mean dbh predictions may also help forest managers decide on a suitable rotation length and thinning regime. However, it should be noted that mean dbh predictions are most useful in stands with normal (bell-shaped) stem size distributions, usually conifers or fast-growing broadleaves. Slower growing broadleaf species such as mahogany rarely show a normal distribution. In Martinique, Tillier (1995) observed in even-aged mahogany plantations that only the dominant trees had a normal distribution, while dominated stems showed an irregular distribution. Thus mean dbh figures for mahogany stands may hide considerable variation, limiting their value as a guide for stand management.

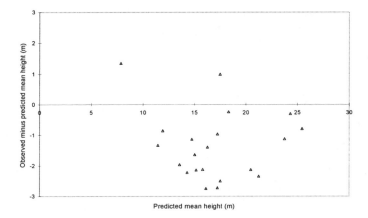

Fig. 9.7. Residual analysis of the relationship between mean height and dominant height (see Table 9.4 for equation). Analysis was performed using independent data from Guadeloupe.

Relationships between mean dbh and age have been reported for some young mahogany plantations (see Table 9.5). The functions give predictions of mean dbh ranging from approximately 10 to 20 cm after 10 years. The wide variations in predictions are probably due to differences in site quality and management techniques between countries. For example, mahogany has shown slow early growth in some line plantings (Oliver 1992), but in unshaded planting sites, Marshall (1939) notes that growth rates immediately after planting out are rapid. The general applicability of the functions appears to be limited and will decline further for closed canopy stands because no stocking density variable has been included.

Table 9.5. Relationships between mean dbh and age

Growth function	Country	Region	Range	R	Source
$D_m = 10^{(1.0649+0.5092\log A - 2.71A^{-1})}$	Fiji	Viti Levu	1-19 yr	2	JICA (1982)
$D_m = 1.49A + 0.09$	Mexico	Tabasco	1-7 yr	3	Llera & Melandez (1989)
*$D_m = 0.0104A^3 - 0.0467A^2 + 1.1646A$	Vanuatu	Many Isles	1-15 yr	2	Neil (1986)

See Table 9.1 for other abbreviations R = reliability class
* Function fitted to a hand-drawn curve

Predictions of mean dbh from age for older mahogany stands have been taken from yield tables (see Section 9.2.3). The dbh-age relationships shown are associated with specific thinning regimes and are given for different site classes (see Figs. 9.8a-e). Yield tables are not accompanied by dbh growth functions and the curves shown are based on interpolation between discrete mean dbh predictions. The predictions will only be applicable to other plantations if prescribed stocking densities are similar throughout the rotation.

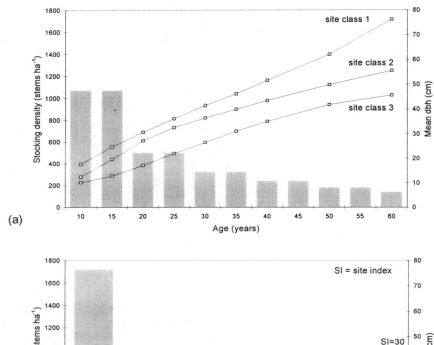

(a)

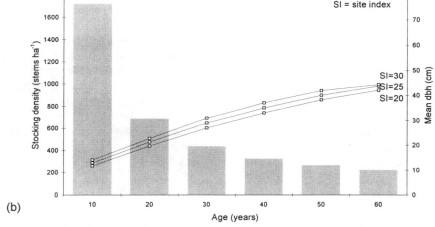

(b)

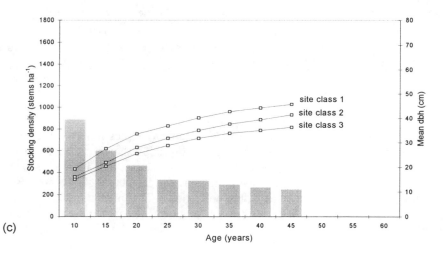

(c)

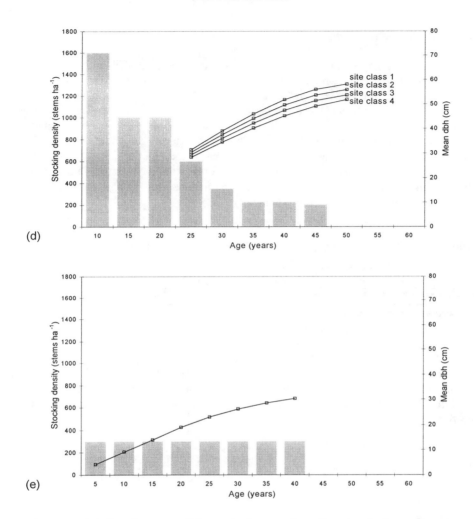

Fig. 9.8. Predicted mean dbh against age (by site class) for mahogany plantations in: (a) Indonesia, (b) Philippines, (c) St. Lucia, (d) Martinique and (e) Belize. Predictions are taken from yield tables (see Appendix V) and are valid for given thinning regimes only.

The curves predict a mean dbh of 25-45 cm (depending on stocking density and site quality) at 30 years, from which point increment begins to tail off. A number of authors have described similar patterns of diameter increment. In Martinique, maximum diameter increment occurs in the first 25 years (Chabod in press). On good sites, growth rates of 2 cm yr^{-1} or more over the first 14 years have been recorded in Ecuador and Peru (Flinta 1960) and with regular fertilisation a mean growth rate of up 3 cm yr^{-1} can be sustained over a 15-20 year period (Rodriguez 1996). No such growth rates have been recorded in older stands.

For comparison, readers are referred to Section 9.1.2 (Figs. 9.3 and 9.4) for mean dbh-age data from other countries and associated functions.

9.2.2.2 Dominant diameter

Dominant dbh may be more useful to foresters than mean dbh, since it reflects more precisely the dbh of trees which will ultimately produce the final crop. Dominant dbh increment may be up to 3 cm yr^{-1} over the first 15 years of growth (Ponce 1933). However, only Tillier (1995) has attempted to produce a dominant diameter growth function (see Table 9.6). The function includes stocking as an index of competition between trees, and dominant height to explain any variation in dominant diameter due to variation in site quality. Age was also included because the function failed to fit observed dominant diameter values satisfactorily using only dominant height and stocking. Although improving the quality of prediction, the inclusion of the age variable makes it impossible to show the relationship graphically.

Table 9.6. Relationship between dominant diameter and dominant height for mahogany

Growth function	Country	Region	R	Source
$D_d = \pi^{-1} e^{(0.01A + 0.018Hd - 0.0045N/Hd + 4.16)}$	Martinique	Pitons du Carbet	1	Tillier (1995)
See Table 9.1 for other abbreviations				R = reliability class

The function was validated using data from Guadeloupe. Residual analysis (see Fig. 9.9) indicates that predictions of dominant diameter are crude, despite the inclusion of variables of dominant height, stocking and age. The function is therefore unlikely to be useful for predictions of dominant height in plantations outside Martinique without modification.

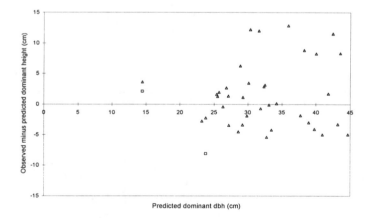

Fig. 9.9. Residual analysis of the relationship between dominant diameter and dominant height (see Table 9.6 for equation). Analysis was performed using independent data from Guadeloupe and Honduras.

9.2.3 Yield predictions

Yield tables may be used to predict timber volumes and the relationships between a range of stand parameters over the course of the rotation. Six different yield tables for mahogany plantations, predicting standing timber volume and associated stand parameters, are reported in the literature and have been reproduced here (see Table 9.7). The yield tables were accompanied by sufficient information to permit an assessment of their reliability (see notes accompanying Tables V-A to V-F in Appendix V). Few of the yield tables have associated growth functions (see notes accompanying Tables V-A to V-F for available details).

According to the classification of Somner and Dow (1978), the mahogany yield tables fall into one of the following three categories:

- Provisional yield tables (produced where permanent sample plot (PSP) data is available for part of the rotation, but extrapolation or use of temporary sample plot data is required to make predictions for older stands).
- Normal yield tables (produced where the PSP data is available for most or all of the rotation; independent variables are stand age and site index, with one or more dependent variables; it is assumed that stands are fully stocked throughout the rotation).
- Empirical yield tables (similar to normal yield tables, but based on plots of average, rather than full stocking).

The yield tables have been classified accordingly (see Table 9.7). The more advanced forms of yield prediction, including variable density yield tables, yield regulation and management tables and growth simulation models (Somner and Dow 1978) have not been produced for mahogany stands.

Table 9.7. Attributes of mahogany plantation yield tables (see Appendix V, Tables V-A to V-F for actual tables)

Country	Yield table parameters	Category	Source
Belize	Age, H_d, D_m, G, V_{10}, V_{MAI}	Empirical	Ennion (1996)
Fiji	Age, V_{MAI}, V	Provisional	AIDAB (1986)
Indonesia	Age, S, N, D_m, V_{MAI}	Provisional	Anon. (1975)
Martinique	Age, H_d, N, H_m, D_m, G, n, v_m, v_{20}, d_g, V_7, V_{20}, V_m, V_{MAI}, V_{CAI}	Normal	Tillier (1995)
Philippines	Age, S, N, D_m, H_d, V_{20u}	Normal	Revilla et al. (1976a)
St. Lucia	Age, S, N, G, D_m, $V_{7.6u}$	Empirical / provisional	Whitman (no date)

The main factors which govern plantation yield are the productive capacity of the site, the stocking of the stand and the silvicultural treatment given to the crop (Pandey 1987). To give accurate yield predictions, yield tables need to address all these factors. Most mahogany yield tables include predictions for different site classes and stocking density. However, few tables describe yield from thinnings (see Appendix V, Tables V-E and V-F for the only reported examples). There are a number of possible reasons for this omission, including: the lack of thinning (forest managers rely on self-thinning through natural mortality); the lack of a local market for thinnings (thinnings are of no merchantable value); and the lack of data on which to base any prediction of thinning yield.

Mahogany yield tables from St. Lucia and the Philippines (see Appendix V, Tables V-A and V-D) which omit thinning volume and predict standing timber volume rather than total volume, do not give a good indication of stand productivity and are therefore limited in their applicability. Yield tables for plantations in Indonesia (see Table V-B), Belize (see Table V-C) and Martinique (see Table V-F) contain details of total volume and volume MAI. The yield table from Martinique is the most comprehensive reported and is based on measurements taken over the course of a whole rotation. The predictions are made for four different site classes and all stand parameters of use to the forest manager have been included. However, whether this or any other yield table is applicable to plantations in other countries will depend on the feasibility of matching site classes and prescribed stocking densities (thinning regimes).

Pure mahogany plantations in Indonesia reach a maximum MAI of 18 m^3 ha^{-1} yr^{-1} in 20 years on the best sites, producing up to 900 m^3 ha^{-1} over the rotation (Figs. 9.10a and 9.11a). On poorer sites, total volume production is 750 m^3 ha^{-1} and the maximum MAI of 13 m^3 ha^{-1} yr^{-1} may not be acheived for 50 years. Plantations in Martinique produce 600-1100 m^3 ha^{-1} over the rotation depending on site quality and reach a maximum MAI in 35 years regardless of site quality (Figs. 9.10b and 9.11b). A maximum MAI of over 30 m^3 ha^{-1} yr^{-1} has been recorded on the best sites (Tillier 1995), although figures in the range 14-25 m^3 ha^{-1} yr^{-1} are more usual. Plantations in neighbouring Guadeloupe are less productive, with the best sites giving 17 m^3 ha^{-1} yr^{-1} over the first 25 years (Soubieux 1983).

In contrast, a total production of around 280 m^3 ha^{-1} and a maximum MAI of 6 m^3 ha^{-1} yr^{-1} over a 50 year rotation is anticipated for plantations in Belize (see Figs. 9.10c and 9.11c). The variation in plantation productivity between countries is explained by variations in stocking density. In Belize, a constant density 300 stems ha^{-1} throughout the rotation is considerably lower than densities of thinned stands in Indonesia and Martinique. Conversion line plantings in Fiji, with a stocking density similar to those in Belize, currently have a comparable MAI of 5 m^3 ha^{-1} yr^{-1} (Singh pers. comm.). However, variations in site quality and differences in the volume concept (e.g. varying top diameters) limit the comparability of total volume and MAI figures.

For comparison, readers are referred to Section 9.1.2 (Fig. 9.5, derived from Appendix III) for volume MAI-age data from other countries and associated functions. It appears that yield tables produced for Indonesia and Martinique predict MAI figures considerably higher than MAIs observed in most other countries. However, MAI figures from other countries are generally taken from stands of lower density.

9.3 Estimation of single tree volume

The volumes given in yield tables (see Section 9.2.3) represent the volume of a stand of trees (measured in m^3 ha^{-1}). The estimation of the volume of a *single tree* (measured in m^3) is a necessary first step to the estimation of the volume of all trees in a stand. Functions have been produced for mahogany which relate volume to dbh (see Table 9.8a) and volume to dbh and height together (see Table 9.8b). Many of the yield tables already described have used these single tree volume functions in the calculation of stand volume.

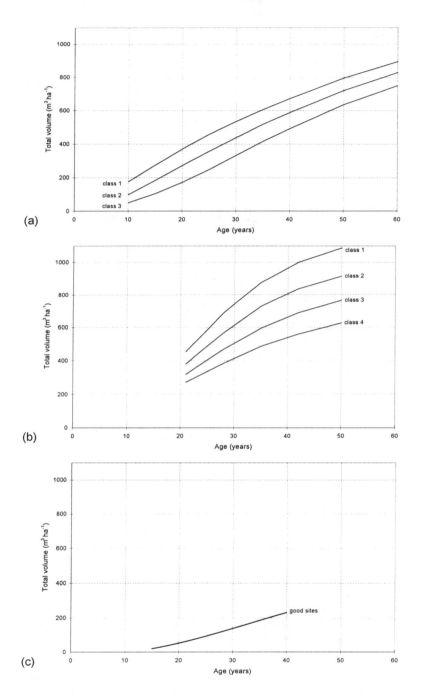

Fig. 9.10. Total volume production against age (by site class) for: (a) Indonesia (to unknown top diameter), (b) Martinique (to 7 cm top diameter) and (c) Belize (to 10 cm top diameter). Predictions valid only for associated thinning regime (see Appendix V for details).

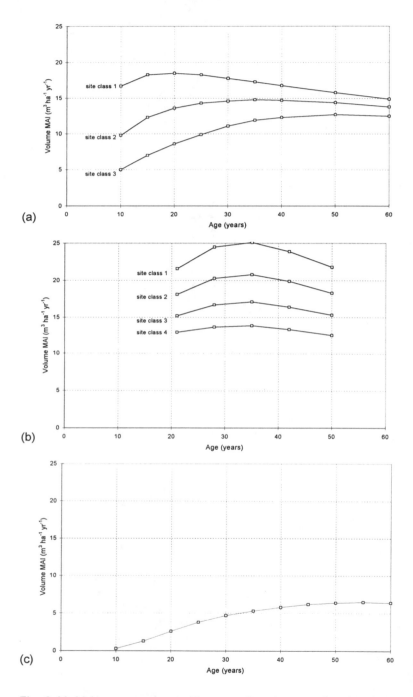

Fig. 9.11. Volume mean annual increment against age (by site class) for: (a) Indonesia (to unknown top diameter), (b) Martinique (to 7 cm top diameter) and (c) Belize (to 10 cm top diameter). Predictions valid only for associated thinning regime (see Appendix V for details).

Table 9.8a. Relationship between single tree volume and dbh for mahogany

Country	Honduras	Sri Lanka	Fiji	Philippines
Region	Lancetilla Gardens	Kurunegala and Kegalle districts	Nukurua	Los Banos
Source	Funes et al. (1983)	Mayhew (in press, b)	Wescom (1979)	Orden (1956)
Equation	$V = 2.125658 \times 10^{-5} (D^{2.6646})$ (logarithmic)	$V = 0.056 - 0.01421D + 0.001036D^2$	For good-average sites: $V = -0.0174 + 4.8470 \times 10^{-4} (D^2)$ For poor sites: $V = -0.0278 + 4.2274 \times 10^{-4} (D^2)$	*$V = 0.0005D^2 - 0.0085D + 0.104$
Definition of estimated volume	Main stem to 1st branch or top diameter, underbark	Main stem to top diameter, regardless of branches, overbark	Main stem to 1st branch or top diameter, overbark	Main stem to 1st branch or top diameter, underbark
Top diameter	30 cm (?)	10 cm, or to last straight 2 m log	10 cm	15 cm (?)
Sample size	195 trees	319 trees	Unspecified	250 trees
Data range	dbh 30 cm +, no trees with a 1st branch height of <2.5 m included	dbh 10-90 cm, few sample trees over 70 cm	dbh 10-44 cm	dbh unspecified, trees 18-38 years old
Mensuration	Measurements taken with a relascope every 3 m up the main stem	Measurements taken with a relascope every 2 m up the main stem	Unspecified	Unspecified
Reliability class	2	1	2	2

See Table 9.1 for abbreviations

* Function fitted to a hand-drawn curve

Table 9.8b. Relationship between single tree volume and dbh with height, for mahogany

Country	Honduras	St. Lucia	Sri Lanka
Region	Lancetilla Gardens	Island-wide	Kurunegala and Kegalle districts
Source	Funes et al. (1983)	Andrew (1994)	Mayhew (in press)
Equation	$V = 1.13036 \times 10^{-4} (D^{1.9688} H^{0.6889})$ (logarithmic)	$V = 0.0260 + 2.71503 \times 10^{-5} (D^2 H)$	$V = -0.0050 + 0.000127 D^2 - 0.001854 H + 0.000025 D^2 H$
Definition of estimated volume	Main stem to 1st branch or top diameter, underbark	Main stem to top diameter, regardless of branches, underbark	Main stem to top diameter, regardless of branches, overbark
Top diameter	30 cm (?)	9 cm	10 cm
Definition of measured height	Merchantable height (i.e. to first branch or top diameter)	Total height	Total height
Sample size	195 trees	26 trees	319 trees
Data range	dbh 30 cm +; height range unspecified, no trees with a 1st branch height of <2.5 m included	height and dbh range unspecified; two trees (one dominant and one average) felled in 13 plots	dbh 10-90 cm, though few samples over 70 cm; height 9-39 m, though few trees over 35 m
Mensuration	With relascope every 3 m up the main stem	With tape every 1 m along main stem	With relascope every 2 m up the main stem
Reliability class	2	3	1

Country	Philippines	Philippines	Guadeloupe
Region	Many islands	Many islands	Basse Terre
Source	Revilla (1976b)	Revilla (1976b)	Soubieux (1983)
Equation	$V = 8.8070 \times 10^{-5} (D^{2.0194} H^{0.7027})$ (logarithmic)	$V = 9.2117 \times 10^{-5} (D^{1.9362} H^{0.7760})$ (logarithmic)	$V = -1.610 \times 10^{-2} + 5.2321 \times 10^{-3} (\pi D) + 8.2786 \times 10^{-6} (\pi D)^2$ $-1.1053 \times 10^{-5} (\pi D) H^2 + 2.9482 \times 10^{-6} (\pi D)^2 H$ $-3.7834 \times 10^{-3} (\sqrt{((\pi D)^2 H)}) + 8.0387 \times 10^{-4} (\pi D) H$
Definition of estimated volume	Main stem to 1st branch or top diameter, underbark	Main stem to top diameter, regardless of branches, underbark	Main stem to top diameter, regardless of branches, overbark
Top diameter	20 cm	20 cm	7 cm
Definition of measured height	Merchantable height (i.e. to first branch or top diameter)	Merchantable height (i.e. to top diameter)	Total height
Sample size	123 trees	123 trees	925 trees
Data range	20-100 cm; 2-30 m	20-100 cm; 2-30 m	12-37 cm; 9-25 m
Mensuration	With relascope every 2.5 m up the main stem	With relascope every 2.5 m up the stem	Unspecified
Reliability class	1	1	1

See Table 9.1 for abbreviations

The single tree volume functions have been used to produce single and double entry volume tables (see Appendix VI). It should be noted that different functions use different top diameter specifications, depending on merchantable log dimensions (see Tables 9.8a and 9.8b). The top diameter may be defined as the diameter at the first branch (variable), the diameter at the top of the last straight log (variable), a fixed diameter (e.g. 10 cm) or a combination of these specifications. Volume estimations may be given under- or overbark. In double entry volume tables, specified heights may be merchantable height or total height.

The variation between double entry volume tables (see Appendix VI, Tables VI-B to VI-G) is due primarily to differences in authors' specifications. Where the specifications are similar, as for volume tables from St. Lucia, Sri Lanka and Guadeloupe, there is little variation in single tree volume predictions (see Table 9.9). In contrast, the variation between single entry volume tables is a result of the difference in authors' specifications *and* the variation in the mean height of trees on which the tables were based. As would be expected, variations between volume predictions from the different sources are substantial (see Fig. 9.12).

Table 9.9. Comparison of abbreviated double entry volume tables from St. Lucia, Sri Lanka and Guadeloupe (see Appendix VI for full tables)

dbh (cm)	Total height = 15 m			Total height = 25 m			Total height = 35 m		
	St. Luc.	Sri Lan.	Guad'pe	St. Luc.	Sri Lan.	Guad'pe	St. Luc.	Sri Lan.	Guad'pe
20	0.19	0.17	0.20	0.30	0.25	0.28	0.41	0.33	-
30	0.39	0.42	0.46	0.64	0.63	0.67	0.88	0.83	-
40	0.68	0.77	0.83	1.11	1.15	1.22	1.55	1.53	-
50	1.04	-	-	1.72	1.83	-	2.40	2.44	-
60	1.49	-	-	2.47	2.66	-	3.45	3.54	-

units = m^3

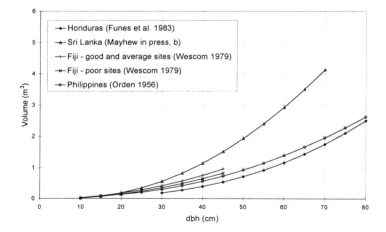

Fig. 9.12. Single tree volume against tree dbh for plantations in Honduras, Sri Lanka, Philippines and Fiji. Volume predictions from single entry (dbh only) volume equations (see Table 9.8a for details).

If any of the volume functions or tables described are applied to plantations other than those for which they were produced, it is important to ensure that specifications are compatible, or serious errors in volume predictions may arise.

9.4 Summary

- Available growth data have been used to produce empirical relationships between: dominant height and age; mean height and dominant height; mean dbh and age (by stocking density); mean dbh and age (by country); and volume mean annual increment and age (by stocking density). The general applicability of the relationships is greatly compromised by the poor quality of the data. The relationships serve to illustrate the high variability in the growth rates of mahogany plantations in different countries.
- Available growth functions describe empirical relationships between: mean height and age; mean height and mean dbh; dominant height and age; mean height and dominant height; mean dbh and age; and dominant dbh and dominant height. No confidence limits are associated with the functions. Relationships between dominant height and age are the most reliable, because they have been based on reasonably large samples.
- Relationships between dominant height and age have been used to define sets of site index curves. Forest managers unable to derive local site index curves should be able to select a function (from Table 9.3) which is compatible with the pattern of growth observed in local plantations. New site index values can be substituted into the function to obtain a spread of curves suitable for local sites.
- Yield tables are available for both high density, pure mahogany plantations (St. Lucia, Indonesia, the Philippines and Martinique) and low density, line plantations (Belize, Fiji). The yield table from Martinique (see Appendix V, Table V-F) is the most reliable and comprehensive, describing parameters for both standing trees and thinnings. Application of one of these yield tables to mahogany plantations in a different country may be possible, but it is important to match site quality, initial stocking density, thinning regime and volume specification.
- The yield tables provide evidence that mahogany plantations can be relatively productive (for a hardwood species). If managed correctly, the maximum mean annual increment for densely stocked plantations ranges between 10 and 25 m^3 ha^{-1} yr^{-1}, depending on site quality. Low density conversion line plantings have a mean annual increment of only 4 to 8 m^3 ha^{-1} yr^{-1}.
- Single entry (dbh) volume tables from Honduras, Sri Lanka, Fiji and the Philippines and double entry (dbh and height) volume tables from Honduras, St. Lucia, Sri Lanka, the Philippines and Guadeloupe are available for the calculation of single tree volume. Forest managers unable to derive local volume tables should be able to select a table (see Appendix VI) which is compatible with single tree volumes observed in local plantations. However, careful attention should be paid to associated top diameter specifications, which may not be compatible with local specifications.
- If forest managers intend to adopt any of the growth functions, yield tables or volume tables described, *it is strongly recommended that they are validated using any data that are locally available.* If no such data are available, it is advisable to match local

conditions of climate and site as closely as possible to those in the plantations from which the growth curves were derived (see Chapter 6 for relevant information).
- Extrapolation outside the ranges given may produce unrealistic results.

Chapter Ten
Timber Quality

The characteristics of natural mahogany timber are well documented (see Table 10.1). The timber is generally regarded as superior to other species of the Meliaceae, with the exception of *Swietenia mahagoni*. It is highly valued for its small dimensional movement, lack of distortion and good finishing qualities (Building Research Establishment 1972). The density of natural mahogany trees is quite variable, ranging from 550 kg m^{-3} for the lightest (yellow) coloured timber to an exceptional 710 kg m^{-3} for the darkest (red-brown) coloured timber, with mean density of 570 kg m^{-3} for 20 samples from Bolivia (de Irmay 1948). The *sapwood* density of natural mahogany trees in Central America ranges from 570 kg m^{-3} to 680 kg m^{-3}, averaging 620 kg m^{-3} for 69 samples from 14 locations (Boone and Chudnoff 1970).

Table 10.1. Characteristics of natural mahogany timber (Building Research Establishment 1972)

Characteristic	Assessment
Colour	Varies from light reddish- or yellowish-brown to dark reddish-brown, darkening on exposure. The heavier material is generally darker in colour. Has a high degree of natural lustre
Grain	Tends to be interlocked, but there is a reasonable proportion of plain, straight-grained timber. Irregularities of grain produce a variety of figure. Flat-sawn timber shows a growth ring figure. The texture is moderately coarse
Density	540 kg m^{-3} after seasoning, notably lighter (and softer) than *S. mahagoni*
Strength (at 12% moisture content)	Bending strength: 83 N mm^{-2} Modulus of elasticity: 8800 N mm^{-2} Compression parallel to grain: 44.2 N mm^{-2}
Movement	Moisture content in 90% relative humidity: 19% Moisture content in 60% relative humidity: 12.5% Corresponding tangential movement: 1.3% Corresponding radial movement: 1.0%
Shrinkage (green to 12% moisture content)	Tangential: about 3% Radial: about 2%
Durability	Sapwood is prone to attack by powder-post and furniture beetles. Heartwood is extremely durable. Both sapwood and heartwood are extremely resistant to preservative treatment

Timber characteristics of plantation-grown mahogany timber have generally been described in terms of density, colour and grain (see Table 10.2). Research has focused on density, which is typically between 500 to 550 kg m^{-3} (slightly less dense than natural mahogany timber). From an analysis of only six sample trees, Briscoe *et al.* (1963) found that wood density increased with growth rate in both dbh and height, although did not vary with variations in growth rate along different radii on the same cross-section. Within individual trees, the wood immediately surrounding the pith was found to be of lowest density. Density was found to be high at the base of the tree, dropping to a minimum at a height of 2.4 m then increasing upwards to a maximum near the base of the crown. Francis (pers. comm.) found that young trees, with a high proportion of juvenile wood, have lower densities than old trees. In contrast, Chudnoff and Geary (1973) believe that there is little difference in timber density between young and old trees.

Mahogany logs containing sapwood are marketable in some countries, such as the Philippines, where there is high local demand for the timber (Malvar pers. comm.). However, mahogany sapwood is usually unacceptable. In common with most species, fast-growing mahogany trees tend to have a greater sapwood:heartwood ratio than slow-growing trees. The sapwood of mature mahogany trees from Fiji has an average width of 4 cm and is lighter and more yellow in coloration than the heartwood (Bulai 1993). Sapwood in slower-growing Malaysian mahogany was only 2.5 cm in width in a 53 cm dbh tree (Streets 1962). The ratio of sapwood to heartwood for smaller stems has not been described, although Francis (pers. comm.) notes that thinnings from Puerto Rican plantations have little value because of the high proportion of undesirable sapwood. It is thought that the strength of the timber might be adversely affected by a high proportion of sapwood (Chand pers. comm.).

Table 10.2. Some of the characteristics of plantation-grown mahogany timber

Country	Air dry density	Grain	Colour	Source
Fiji	545 kg m^{-3} (845 kg m^{-3} wet, 474 kg m^{-3} basic)	marked areas of interlocking grain; high natural lustre	slightly lighter than natural mahogany; yellow to pale pink, darkening on exposure	Bulai (1993), Chand (pers. comm.)
Honduras	slightly less dense than natural mahogany	nda	nda	Wangaard & Muschler (1952)
Philippines	510-560 kg m^{-3}	nda	nda	ERDB (1996)
Puerto Rico	around 550 kg m^{-3} but as low as 450 kg m^{-3} in young trees	nda	nda	Francis (pers.comm)
Trinidad	denser than natural mahogany	'somewhat wild'	pale at first, darkening on exposure	Lamb (1955)
Western Samoa	density of 540 kg m^{-3} for trees from a 29 year old plantation	interlocking grain	lighter than natural mahogany	Haslett (1986)

nda = no details available

The nature of defects in Fijian plantation timber and their effect on timber grade was examined by Cown *et al.* (1989) and McConchie *et al.* (1989). The biggest problem was the high incidence of rot which was not apparent from the outside of the logs. Of a total of 161 sample logs, 50% were affected (20% had pocket rot and 30% had severe rot). The cause of this rot is not described, although it is noted that knife cuts from the vine cutting operations in Fiji often caused rot to develop around the debarked area. The rot may be associated with termite damage (see Section 12.1.2). The other significant defects were pin-knots (epicormic knots up to 8 mm) and pin-holes caused by ambrosia beetle attack. Both pin-knots and pin-holes tend to be concentrated around the centre of the stem. Of all sawn logs, 34% had some timber downgraded due to pin-holes. This figure would have been higher if the grading rules were not so lenient: at the moment an unlimited number of pin-holes are permitted on the reverse face and edge of select grade timber. Cown *et al.* (1989) conclude that the market may resist substitution of natural for plantation mahogany from Fiji.

10.1 Effects of site and climate on timber quality

In natural forests, there is some evidence that mahogany trees growing in marginal sites produce more desirable timber than trees growing well within their natural range. For example, trees on dry, well-drained, rocky soils in northern Belize were found to have hard and compact wood, whilst those in permanently moist conditions of southern Belize had comparatively soft, straight-grained wood, lacking in figure (Stevenson 1944b, Lamb 1960). In Bolivia, trees at high altitudes (500-1200 m) were slower-growing than trees in lowland forests, but produced better timber (de Irmay 1948).

Plantation timber quality is affected by site and climatic conditions in a similar manner. Mahogany trees on drier sites in Martinique grow more slowly but are always of higher quality than faster-growing trees on wetter sites (Marie 1949). In Kolombangara, Solomon Islands, the absence of a dry season reduces the figure and thus the value of plantation mahogany timber (Isles pers. comm.). In general, conditions found in plantations are much more conducive to rapid growth than those in natural forests and timber quality will almost certainly suffer as a result. However, given the current economic climate, few forest managers are in a position to extend rotation length of mahogany plantations (by planting on poorer sites or at higher densities) in order to improve timber quality. Genetic improvement of planting stock may be the best solution.

10.2 Summary

- The timber of *Swietenia macrophylla* is generally regarded as superior to the timber of other species of the Meliaceae, with the exception of *S. mahagoni*. It is valued for its small dimensional movement, lack of distortion and good finishing qualities.
- Comparisons between natural and plantation-grown timber indicate that plantation-grown timber is slightly less dense and lighter in colour; the grain may be coarser.
- The sapwood of plantation timber, which is unmarketable in most countries, occupies the outer 2.5-5.0 cm of a merchantable (over 50 cm dbh) tree. The ratio of sapwood to heartwood is greater in faster-growing trees.
- Observations from natural forests indicate that plantation timber quality can be improved by reducing growth rates and extending rotation length. It may also be possible to improve timber quality by genetic improvement of planting stock.

Chapter Eleven
Shoot Borer Control

The main factor which has limited the cultivation of mahogany in plantations is attack by shoot boring moths (*Hypsipyla* spp., Lepidoptera, Pyralidae), which are widespread throughout the tropics. The moth larvae destroy the terminal bud of the young tree, reducing vigour and causing forking which lowers the economic value of the timber (see Plates 11.1, 11.2 and 11.3). This pest has resulted in the failure of many attempts at reforestation with mahogany, for example in Puerto Rico, Guatemala, Peru and Cuba (see Newton *et al.* 1993a for details).

The two most important shoot borer species with respect to mahogany are *Hypsipyla grandella* (Zeller) and *H. robusta* (Moore). *H. grandella* is found throughout Central and South America (except Chile), and also occurs on many Caribbean islands and the southern tip of Florida (Entwistle 1967). The closely related *H. robusta* (Moore) is widely distributed throughout West and East Africa, India, Indonesia, Australia and South-East Asia (Entwistle 1967). Available evidence indicates that the shoot borer has not been been able to colonise many Pacific islands to date. *H. robusta* is present in the western and central Solomon Islands (Oliver 1992, Speed pers. comm.) but does not appear to be present in the more remote eastern Solomon Islands, Vanuatu, Fiji or Western Samoa (Neil 1986, Singh pers. comm., Speed pers. comm.).

Bibliographies on *Hypsipyla* have been compiled by Tillmans (1964) and Grijpma and Styles (1973) (see also Whitmore 1976a,b). Both *H. grandella* and *H. robusta* are known to attack *S. macrophylla* and *S. mahagoni* (Newton *et al.* 1993a); less information is available concerning *S. humilis*, although it is certainly attacked by *H. grandella*.

The biology of *H. grandella* is known in some detail (Ramirez-Sanchez 1964, Roovers 1971); less information is available concerning the life history of *H. robusta* (Roberts 1968, Wagner *et al.* 1991) although the two species appear to behave similarly. The total life cycle of both species usually lasts between 1 and 2 months, depending on climate and food availability (Newton *et al.* 1993a). In *H. grandella*, eggs are laid singly, or occasionally in clusters of 3-4, on or near leaf axils, scars or veins. Usually only 1-3 eggs are laid on each tree (Grijpma 1974). The eggs hatch after around 3 days. The newly hatched larvae move towards new shoots and burrow into the stem or leaf midrib, subsequently migrating to the terminal shoot of the stem or main branch. The remainder of the larval stage is spent boring into the primary stem or branches of the tree, feeding on the pith. The combined duration of the 5 or 6 larval stages is around 30 days (Ramirez-Sanchez 1964, Roovers 1971).

Plate 11.3. The effect of shoot borer damage on the form of a mature tree.

Plate 11.2. Shoot borer damage to the leading shoot of a young mahogany tree. Note the presence of frass on this recently damaged shoot.

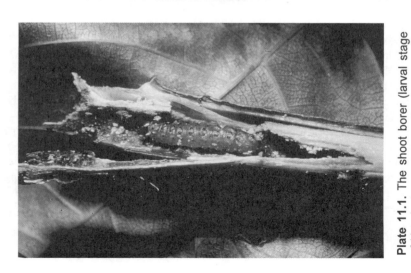

Plate 11.1. The shoot borer (larval stage of *Hypsipyla grandella*) in its tunnel. *Photo:* M. Speight.

As a result of damage to the apical shoot, *Hypsipyla* attack causes loss of form and growth increment, which may be severe, but rarely causes death of the tree (Whitmore 1976a,b, Newton *et al.* 1993a). Attacks tend to be most pronounced on newly produced shoots. Although attack may persist throughout the life of the tree, it is the first few years of growth which are critical in terms of the effect of the pest on timber merchantability. Morgan and Suratmo (1976) found that trees aged 3-6 years and between 2-8 m tall are most susceptible to damage; infestation was about 90% for trees aged 3 years / 2.5 m high (Suratmo 1979).

A number of attempts have been made to use biological, chemical and silvicultural methods for controlling the mahogany shoot borer (Grijpma 1973, 1974, Whitmore 1976a,b), but many of these have failed to reduce damage to economically acceptable levels. A review of the various methods of control that have been attempted is given below (see also Newton *et al.* 1993a).

11.1 Silvicultural control of *Hypsipyla* spp.

Existing evidence for successful silvicultural control of *Hypsipyla* is conflicting and to a large extent anecdotal (Hauxwell *et al.* in press). Attack has traditionally been correlated with factors such as shade, establishment technique, species mixtures and site characteristics. These factors have often been poorly defined and their interdependence greatly hinders interpretation of both anecdotal and scientific observations. However, it is possible to reconcile apparently contradictory evidence for successful control if silvicultural treatments are considered in terms of their effect on the interaction between plant and insect. Following Hauxwell *et al.* (in press), four mechanisms of interaction are listed here (see Section 11.1.1), which explain in precise terms how the damaged caused by the shoot borer may, in theory, be reduced by silvicultural methods. It must be emphasised that each of these mechanisms is hypothetical; there are few scientific data available which may be used specifically to test their validity.

The range of silvicultural techniques that have been suggested for *Hypsipyla* control are also discussed here with reference to these mechanisms. The techniques have been divided into those relevant to plantation planning, i.e. pre-planting (see Section 11.1.2) and those that are relevant to plantation maintenance, i.e. post-planting (see Section 11.1.3). Observations of the success or failure of various silvicultural techniques are recorded (see Tables 11.1a and 11.1b). These observations have been used as evidence for and against various hypotheses which have been stated (often implicitly rather than explicitly) in the literature. It should be recognised that the lack of evidence for a particular hypothesis usually reflects a lack of appropriate research, rather than a refutation. Hypotheses which are not supported by any observations are discussed in the text.

11.1.1 Proposed mechanisms for silvicultural control

The four mechanisms which may account for silvicultural control of shoot borer damage involve: location, susceptibility, tolerance and natural enemies (Hauxwell *et al.* in press). The following sections describe, and provide evidence in support of each mechanism.

11.1.1.1 Location

Mechanism: the *Hypsipyla* moth is prevented from locating the mahogany host.

Physical screening from Hypsipyla. In theory, the moth may be prevented from reaching the mahogany plant by the presence of vegetation (including anything from grass to trees) around individual mahogany trees or possibly around whole stands. The vegetation itself may be a solid barrier, or the microclimate that is associated with a particular vegetation type may act as an atmospheric barrier if it is repellent to the moth (Grijpma and Gara 1970). If the mahogany tree or stand is surrounded by a large area of terrain inhospitable to the moth, the distance between the tree or stand and the nearest shoot borer habitat may also act as a barrier. Although experimental evidence for this mechanism is lacking, Abe (1983) notes that the adult moth has a limited flight ability and is unable to cover large distances.

Chemical screening from Hypsipyla. As *Hypsipyla* moths primarily locate their host trees by olfaction, chemical signals given off by non-Meliaceae species in or around a stand may in theory be sufficiently strong to prevent the moth from locating the mahogany trees (Grijpma and Gara 1970, Grijpma 1976, Morgan and Suratmo 1976, Kareira 1983). A number of species may be *Hypsipyla* repellent, including *Azadirachta indica*, *Quassia amara* and possibly *Erythrina poeppigigiana* (Watt, pers. comm.). *Azadirachta excelsa* may also have potential in this context (Matsumoto, pers. comm.). If their repellence could be demonstrated, these species would make suitable planting associations. However, only anecdotal evidence in support of their use is available at present.

11.1.1.2 Susceptibility

Mechanism: the susceptibility of mahogany to shoot borer attack is reduced.

Natural resistance to Hypsipyla *attack.* Certain provenances of mahogany may be less susceptible to shoot borer attack, either by being less attractive to the insect, or by the possession of toxins which kill the larvae. Newton *et al.* (in press) present preliminary evidence in support of this mechanism, indicating that there may be genetic variation in the susceptibility of mahogany to attack at both the provenance and family level.

Silvicultural measures enhancing resistance to Hypsipyla *attack.* Through site selection and silvicultural manipulation (e.g. the provision of shade) it may be possible to induce changes in shoot composition and thus reduce susceptibility to attack. Evidence for this mechanism is indirect. In Peruvian Amazon line plantings, Yamazaki *et al.* (1992) found that mahogany saplings suffered a large increase in incidence of *Hypsipyla grandella* attack during periods of active shoot growth (flushing). Similar results were obtained by Newton *et al.* (1998) in Costa Rica and by Frederick (1994), whose survey of 491 trees in a 2-year-old plantation in Grenada indicated that larger and thus faster-growing saplings suffered more attack. The tendency of shoot borer attack to be associated with flushing shoots may reflect a change in their chemical composition, which influences larval burrowing and/or oviposition of the adult moth (Grijpma 1976, Vega 1976). It is also possible that the forest environment affects physical plant defences such as sap flow (Lamb 1966) and chemical plant defences such as sap tannin content (Lamb 1966), resin flow (Wilkins 1972, Whitmore 1978, Styles in Pennington 1981) or content of complex secondary plant compounds such as proanthocyanidins or limonoids which may be insecticidal or antifeedants (Adesogan and Taylor 1970, Adesida and Adesogan 1971, Koul and Isman 1992, Vanucci *et al.* 1992). However, little is known about these defences or means of enhancing them.

11.1.1.3 Tolerance

Mechanism: the ability of mahogany to recover after attack is enhanced.

Natural tolerance to Hypsipyla *attack.* Certain provenances of mahogany apparently display a relatively high ability to recover from attack by vigorous apical growth, as a result of higher apical dominance (Newton *et al.* 1996). A decapitation test may be used in the nursery to screen for this characteristic; the approach revealed genetic variation in apical dominance of *Cedrela odorata*, a close relative of *Swietenia* (Newton *et al.* 1995).

Silvicultural measures enhancing tolerance to Hypsipyla *attack.* Silvicultural treatments which promote growth of the leading shoot may enable trees to recover from damage caused by the *Hypsipyla* larva. Trees on sites to which they are well matched grow relatively rapidly and consequently may be better able to recover after attack by production of a vigorous apical shoot (Grijpma 1976, Vega 1976). Pruning the tree to remove all but the strongest and straightest of the multiple leaders may help to improve form. Other silvicultural techniques rely on using lateral shade to encourage strong vertical growth (Stevenson 1939, Aubreville 1953, Yared and Carpenezzi 1981). On open sites, stocking at high density and using a mixture of mahogany with a slightly faster growing species are possible options. On sites with existing natural vegetation, residual vegetation between planting lines may provide useful lateral shade. Retention of regrowth around individual trees may also encourage vertical growth of mahogany trees.

11.1.1.4 Natural enemies

Mechanism: *Hypsipyla* is controlled by natural enemies.

Natural enemies of *Hypsipyla* may help to reduce numbers of adult moths and possibly to remove eggs or larvae from shoots before any damage has been caused. Predators, parasitoids and pathogens may be encouraged by providing a habitat with alternative food sources. Habitats for natural enemies may be created by mixed planting. Tree crop species which are known to harbour large populations of ants (including *Citrus*, mango, *Macadamia*) and cover crops which harbour natural enemies have potential in this context (Watt pers. comm.). Retention of natural vegetation between planting lines may theoretically help to maintain existing habitats for natural enemies (Yared and Carpenezzi 1981), although if any species of Meliaceae are present, the residual natural forest may also serve as a source of *Hypsipyla* populations.

Silvicultural treatments may also change microclimates, affecting the action of natural enemies. For example, increasing humidity improves germination of some entomopathogenic fungi (Ferron 1981), and reducing solar UV improves persistence of entomopathogenic viruses (Entwistle and Evans 1983). The influence of silvicultural interventions on populations of natural enemies of the shoot borer remain largely uninvestigated.

11.1.2 Techniques for silvicultural control pre-planting

11.1.2.1 Provenance selection

Careful provenance selection may help to reduce shoot borer damage to new mahogany plantations (see Sections 11.1.1.2 and 11.1.1.3) but more extensive provenance testing is urgently required. It is likely that the performance of any provenance will be influenced by site characteristics and this interaction with site may, in turn, influence resistance to shoot borer attack. For example, the ability to tolerate attack by vigorous growth of the apical shoot will tend to be more pronounced on sites where overall growth is vigorous (i.e. on sites with high soil fertility, adequate drainage, etc.).

11.1.2.2 Site selection

The idea that trees growing on fertile soils suffer less shoot borer damage than trees on infertile soils has been proposed by several authors (see Table 11.1a). Fertile soils may reduce plant susceptibility and/or increase plant tolerance to *Hypsipyla*. However, some argue that fertile soils produce especially succulent shoots which are attractive to the shoot borer and recommend planting on poorer soils to reduce attack (e.g. Akanbi 1973). Such a strategy may reduce plant susceptibility, but is unlikely to increase tolerance to attack. The presence of certain chemicals, such as calcium, in the soil may also reduce susceptibility and/or increase plant tolerance (see Table 11.1a).

Site exposure may affect shoot borer damage, but there is no consensus on which sites are most at risk (see Table 11.1a). The observations may be explained by the failure of *Hypsipyla* to locate the mahogany host and/or the presence/absence of natural enemies on exposed sites. In fact, there is likely to be a strong correlation between exposure and site quality, which would bring mechanisms of susceptibility and tolerance into play.

It has been suggested that plantations can be isolated from *Hypsipyla* habitats (see Table 11.1a). Spatial isolation may be achieved by planting in sites where *Hypsipyla* is unlikely to be found or creating a barrier around the plantations which is impenetrable by *Hypsipyla*. Failure of *Hypsipyla* to locate the host is the only feasible control mechanism.

11.1.2.3 Plantation design

The design of a plantation, particularly in terms of its composition and layout, is the pre-planting decision over which foresters have most control. The design of a plantation is significantly affected by presence of pre-existing vegetation. Where there is a residual natural forest cover, low density planting will usually be preferred because of the high financial and environmental costs of forest clearance. Where previous forest cover has previously been removed, high density planting will usually be preferred because of higher potential productivity per unit area. Silvicultural techniques used for shoot borer control are therefore categorised by pre-planting conditions. (Readers are referred to Chapter 7 for a detailed discussion of establishment techniques.)

Planting in natural forest. In areas of disturbed or regenerating natural forest, low density planting lines have been favoured because of the belief that the natural vegetation affords protection from the shoot borer (see Table 11.1a). Under these circumstances, many (possibly all) of the proposed mechanisms for control may come into action: the mahogany host may not be located due to the presence of the vegetation between the lines; susceptibility to attack may be reduced owing to shaded conditions reducing shoot attractiveness/palatability; tolerance may be increased owing to lateral shade encouraging vertical growth and self-pruning; and the presence of natural enemies in the vegetation between planting lines may limit *Hypsipyla* populations.

Despite many advocates of low density planting (Beard 1942, Holdridge 1943, Cater 1945), there is little evidence that low density line plantings suffer less shoot borer damage than high density plantations (Newton *et al.* 1993a). Many authors have recommended line planting in the belief that the stands thus created resemble natural mahogany stands. 'All the autecological information related to mahogany points to line planting as the ideal method for its establishment ... widely spaced line plantings simulate the conditions in which mahogany successfully regenerates in the natural forest' (Weaver and Bauer 1986). However, natural stands of mahogany, particularly over the critical years immediately after establishment, may be of relatively high density (see Snook 1993). In addition, some authors are dubious about the screening effect of the natural vegetation between planting lines. Extensive damage to line planted mahogany in Peru indicates that planting lines actually served as *Hypsipyla* 'highways' (Ikeda *et al.* in press, Matsumoto pers. comm.).

There is a widespread belief, originating from experience in the natural forest improvement areas of Belize, that the presence of overhead shade may serve to reduce shoot borer attack (see Table 11.1a). As a result, some plantations have been established under a light shelterwood of natural forest canopy trees, with the undergrowth largely cleared. The most likely control mechanism is reduced susceptibility to attack, produced by reduction in shoot attractiveness and/or palatability. The lack of undergrowth may favour the action of some *Hypsipyla* enemies, particularly birds and bats. However, such conditions may assist location by the moth and result in increased attack (Roberts 1966).

Planting on open sites. There has been considerable interest in using a mixture of species to reduce shoot borer damage. A number of silvicultural techniques have been tried (see Table 11.1a). Planting mahogany under a pre-existing plantation has given some positive results in cases where the non-Meliaceae trees had not closed canopy at the time of planting. Establishment of mahogany with fast-growing species also creates conditions of overhead light and lateral shade. In both cases, many (possibly all) of the proposed mechanisms for control may come into action: host location may be prevented, owing to screening by non-Meliaceae; susceptibility to attack may be reduced owing to shaded conditions reducing shoot attractiveness and/or palatability; tolerance may be increased owing to lateral shade encouraging vertical growth and self-pruning; and the presence of natural enemies associated with non-Meliaceae trees may limit *Hypsipyla* populations.

Recently there has been interest in using particular species to protect open planted mahogany, but little evidence to indicate that these species are successful. *Citrus*, with associated colonies of ants (which are possible predators of *Hypsipyla*), has been planted amongst mahogany with varied results (see Table 11.1a). The use of *Hypsipyla*-repellent species (adversely affecting host location) has yet to be reported.

Some authors (e.g. Lapis 1995) recommend high density stocking on open sites to reduce the development of lateral branches caused by the shoot borer (see also Section 11.1.3.4 below). On its own, dense stocking may increase the tolerance of mahogany trees to attack, but it will not trigger any other control mechanisms.

Table 11.1a. Shoot borer control hypotheses to be considered pre-planting

Hypothesis	Evidence for and against	Source
1. Provenance selection		
Certain genotypes display superior resistance to shoot borer attack	**For:** Preliminary evidence available from growth trials in Costa Rica and Trinidad indicates genetic variation at the provenance and family level in susceptibility to attack	Newton et al. (1996)
2. Site selection		
Trees on fertile soils are damaged less than trees on poor soils	**For:** Observations in plantations of Martinique, Puerto Rico and Malaysia	Marrero (1942), Marie (1949), Huguet & Marie (1951), Streets (1962)
Trees on calcareous soils are less damaged than trees on non-calcareous soils	**For:** In Costa Rican research trials, the calcium content of soils is related to the severity and/or incidence of attack on *Cedrela odorata*	Newton et al. (in press)
	For: In plantations on calcareous soils in Belize, *Swietenia macrophylla* suffered less attack	Cree (1953)
Trees in sheltered sites are damaged less than trees in exposed sites	**For:** Shoot borer attack in Martinique and St. Lucia was less severe on sheltered sites and more severe on sites exposed to winds	Marie (1949), James (pers. comm.)
	Against: There is significant decrease in the incidence of shoot borer infestation in mahogany on ridges in Puerto Rico	Weaver & Bauer (1986)
Trees isolated from natural forest are not damaged	**For:** A small mahogany plantation in Belize containing 2-3m tall trees, isolated from natural forest by extensive banana plantations, has not been attacked	Ennion (pers. comm.)
	For: Plots of mahogany at Benekat, Sumatra, surrounded by *Acacia mangium* were not attacked while the *A. mangium* was left unweeded	Matsumoto (pers. comm.)
3a. Plantation design in natural forests		
Trees in planting lines or enrichment lines are damaged less than trees in open plantations	**For:** The shoot borer was virtually absent in planting lines through natural forest in Brazil over a 4 year period	Yared & Carpenezzi (1981)
	For: Planting experience in the Solomon Islands indicates that widely spaced planting lines cut through the bush ensure a low level of shoot borer attack on mahogany (around 10%)	Wilson (1986)
	Against: Three line planted plots of mahogany in Costa Rica suffered serious borer damage and showed no sign of recovery. Lines were cut in 7-8 m high secondary regrowth on abandoned fields	Somarriba (pers. comm.)
	Against: In Peru, line plantings were heavily attacked by the shoot borer in the wet season while young trees were flushing	Yamazaki et al. (1992), Ikeda et al. (in press)
Trees with overhead shade are damaged less than trees with no overhead shade	**For:** In Belize, mahogany raised under a shelterwood of balsa were free of shoot borer infestation	Frith (1958)
	For: Removal of overhead cover in improved natural forest in Belize led to severe shoot borer damage of regenerating seedlings. In contrast, seedlings stagnating in the dense shade of untreated bush were not attacked	Oliphant (1926), Stevenson (1927c), Nelson Smith (1942)

Trees planted in a pre-existing plantation are damaged less than trees planted in the open	**For:** In the Solomon Islands, mahogany was found to grow best in conditions of light overhead shade	Anon. (1979)
	For: Forest managers in Sri Lanka observed that trees planted in open conditions were heavily attacked, but 'under shelter mahogany is entirely free from the shoot borer'	Perera (1955) Ponce (1933)
	For: In Java, attack by the shoot borer was reduced by planting mahogany under a shelterwood	
	Against: The results of an experiment involving the measurement of 1500 line planted saplings ranging from 1-7 years old in Puerto Rico show that the more shaded crowns are just as susceptible to shoot borer attack as the less shaded crowns	Bauer (1987a)
	Against: In both Puerto Rico and Honduras, the presence of overhead shade was not successful in preventing shoot borer attack in line plantings	Whitmore (1976a,b), Chable (1967)

3b. Plantation design on open sites

Trees planted in a mixture with a fast-growing nurse spp. are damaged less than pure plantations	**For:** Mahogany trees planted in young (3-10-year-old) *Artocarpus integrifolia* in Sri Lanka were badly attacked initially, but after a few years the stems showed hardly a trace of the original poor form	McNeil (1935), Perera (1955)
	For: A small plot of mahogany planted in mid-rotation *Pinus caribaea* was unaffected by shoot borer	Weerawardene (1996)
	For: Mahogany trees planted in *Trema orientalis* were protected from the shoot borer	Streets (1962) Iputu (pers. comm.)
	For: A mixed experimental planting of mahogany in young *Securinega flexuosa* in the Solomon Islands produced mahogany with better form than an adjacent pure mahogany control	
	Against: In Costa Rica, a high incidence of *Hypsipyla* attack was observed in trial plots of *Swietenia* planted under *Gmelina arborea* and *Cassia simea*	Combe & Gewald (1979)
	For: In St. Edmund Forest, St. Lucia, a 30-year-old block of mixed mahogany and *Hibiscus elatus* suffered less shoot borer attack than an adjacent block of pure mahogany	Bobb (pers. comm.)
	For: Young mahogany screened by *Inga edulis* suffered less insect attack in Western Brazil	Matos et al. (in press)
	For: In Guatemala, single lines of mahogany were separated by 3 lines of *Cordia alliodora*. After 7 years, the proportion of trees that were attacked was less than 10%	
Trees planted with *Citrus* are damaged less than trees planted without *Citrus*	**For:** In the Sierra de Esparanza mountain range of Honduras, a 3-4-year-old mahogany plantation grown with *Citrus* sp. was observed to be free of shoot borer damage	Meza (pers. comm.)
	Against: Ejido farmers in Quintana Roo, Mexico, report that a mixture of *Citrus* and mahogany has not affected the incidence of shoot borer damage	Hassan (pers. comm.)
Trees planted with other Meliaceae spp. are damaged less than trees planted without other Meliaceae spp.	**Against:** In Sabah, the shoot borer was found to attack mahogany more than *Cedrela odorata*, which in turn was attacked more than *Khaya ivorensis*	Nasi & Monteuuis (1993)
	Against: In adjacent trial plots on Mount Austen, Solomon Islands, *Toona ciliata* was 'virtually destroyed' and *Swietenia macrophylla* 'severely distorted' but *Cedrela odorata* was unaffected	Wilson (1986)
Natural regeneration under mature trees is damaged less than young plantations in open conditions	**For:** Observations of natural regeneration under mahogany plantations in Sri Lanka, Puerto Rico, St. Lucia and the Philippines revealed few signs of shoot borer damage	Mayhew et al. (in press)

Table 11.1b. Shoot borer control hypotheses to be considered post-planting

Hypothesis	Evidence for and against	Source
1. Weeding		
Trees surrounded by tall weeds and/or secondary regeneration are damaged less than trees that have been weeded	For: The shoot borer was controlled in Martinique by maintaining brush, containing fast-growing native species *Cecropia peltata* and *Sterculia caribaea*, between lines. Only vines were eliminated	Marie (1949)
	For: In the Columbia River taungyas of Belize the high incidence of shoot borer attack (up to 61.3%) was explained by rapid growth of the seedlings beyond competing vegetation. There was little shoot borer damage when grass and vegetation surrounding mahogany trees was left to grow high	Kinloch (1933), Villafranco (pers. comm.)
	For: 'Over-weeded' plantations in Silk Grass Reserve of Belize let in 'too much light, thus favouring shoot borer and the development of herbaceous weeds and grass.' Mahogany which had been overtopped by the secondary regeneration suffered less shoot borer damage	Cree (1954)
	For: Shoot borer attack in the Solomon Islands is less severe if crowns are kept close against inter-row vegetation after the first 2 years	Anon. (1979)
2. Pruning		
Trees that have been pruned are more likely to recover than trees that are left with multiple leaders	For: Mahogany responds very favourably to pruning. Regular pruning of trees damaged by the shoot borer in Honduran plantations substantially improved tree form	Chable (1967)
	For: Regular pruning of mahogany plantations in Kolombangara, Solomon Islands, over the first few years of growth produces straight stems, despite recurrent shoot borer attack	Speed (pers. comm.)
	For: In Martinique, most mahogany stems are attacked by the shoot borer, but pruning has ensured the production of 150 high quality timber trees per hectare	Chabod (in press)
3. Fertilising		
Trees that are fertilised are damaged less than trees which are not fertilised	For: In St. Lucia, rapid growth through the application of fertiliser prohibits *Hypsipyla* attack	Vidal (pers. comm.)
	For: Vigorously growing trees on fertilised sites in Malaysia and the Philippines suffer little damage	Streets (1962)
	Against: A mahogany agroforestry system in Brazil with 2 different fertiliser inputs (NPK and P only) suffered a similar incidence of attack (73% vs 81%) after 30 months	Matos *et al.* (in press)
4. Thinning		
Trees in stands that are maintained at high densities are more likely to recover than trees in thinned or low density stands	For: In plantations in Belize, smaller branches of mahogany stems died quickly after canopy closure	Stevenson (1939)
	For: An unpruned, densely stocked stand of mahogany in Quintana Roo, Mexico, was found to contain trees of similar form to those in an adjacent stand that had been thinned and pruned	Parraguirre (pers. comm.)
	For: In St. Lucia, a mixed plantation of mahogany and *Hibiscus elatus* maintained at original (1.5 x 1.5 m) stocking density for 30 years contains mahogany trees of good form	Bobb (pers. comm.)

11.1.3 Techniques for silvicultural control post-planting

11.1.3.1 Weeding and cleaning

Several authors have described the positive effect of maintaining tall weeds and secondary regrowth around individual mahogany trees to reduce shoot borer attack (see Table 11.1b). The presence of weeds may enhance the action of all four proposed mechanisms (failure of host location, reduction in susceptibility, increase in tolerance and increase in natural enemies).

The use of taungya has been recommended as a means of reducing shoot borer attack (e.g. Roberts 1966), largely on the basis of experience with taungyas in Belize. In fact, agricultural crops were usually grown with mahogany over the first year only and therefore provided short-lived protection. After the crops were harvested, the low density of mahogany encouraged the development of weeds and regrowth, which were carefully manipulated to reduce shoot borer damage (Stevenson 1944b). The use of shade-giving weeds and regrowth to reduce shoot borer damage was effective (see Table 11.2) but a balance had to be struck in order to obtain satisfactory mahogany growth rates.

Table 11.2. *Hypsipyla* infestation of young mahogany under different conditions of shade in Belize taungyas (Stevenson 1936)

Plot no.	% of stems attacked		
	Full light	Half shade	Full shade
1928a	-	93	-
1928b	100	0	0
1929a	-	100	-
1929b	-	100	0
1929c	100	29	0
1929d	-	100	15
1930a	100	83	0
1930b	100	0	0
1932a	100	75	0

11.1.3.2 Fertilising

Fertilising mahogany trees may reduce susceptibility to shoot borer damage by altering the chemical composition of the apical shoots and/or enhancing tolerance by improving vigour. Evidence gathered so far is equivocal (see Table 11.1b). Lapis (1995) recommends that fertiliser is not applied in order to avoid production of 'juicy and succulent terminal shoots', but despite this comment, there is no clear evidence that shoots of fertilised plants are actually more susceptible to attack.

11.1.3.3 Pruning

Pruning is perhaps the most direct silvicultural technique for reducing shoot borer related damage and results so far have been positive (see Table 11.1b). Rather than concentrating on removing shoots which have been attacked, pruning should seek to encourage those shoots which are displaying vigorous vertical growth, by removal of those which are not. Improving form by pruning is an artificial means of increasing tolerance to shoot borer attack.

11.1.3.4 Thinning

There is some evidence to indicate that maintenance of high stand density over a number of years through delayed thinning may improve stem form (see Table 11.1b). In pure stands, delayed thinning can only reduce the effects of shoot borer attack by increasing plant tolerance. In mixed stands, the presence of non-Meliaceae species may also limit additional attack through the other control mechanisms.

If mahogany stocking densities are high initially, foresters have the option of thinning out the trees which fail to recover, whilst retaining trees with acceptable form for the final crop. This strategy has been used with success in Martinique, where mahogany plantations are in their second rotation (Chabod in press) and has been recommended for plantations in the Solomon Islands (Sandom 1978). Selective removal of damaged mahogany may help to ensure that seed produced by the final crop is genetically less susceptible or more tolerant to attack.

11.1.4 Evaluation of reported experiences and recommendations

Regional variations in both shoot borer and mahogany populations, and the possible effects of site on pest-host interactions, mean that silvicultural techniques which have been successfully used on one site may be ineffective on another. Nevertheless, an attempt has been made to evaluate the silvicultural techniques used to control the shoot borer (on the basis of evidence in Sections 11.1.2 and 11.1.3), in order to identify those approaches which hold most promise.

Where mahogany is planted in areas where the shoot borer is present, forest managers should assume that the majority of trees in the plantation will be attacked. Although many different silvicultural techniques have been adopted, none have proved to be universally successful in reducing the incidence of attack to acceptable levels. Instead, it is preferable to focus on those silvicultural techniques which encourage recovery after attack, by promoting vigorous vertical growth and restricting the development of lateral branches. This can be achieved by:

- *Careful site matching.* Unhealthy plants have little chance of making a successful recovery after shoot borer attack. Planting sites which are carefully selected to match the requirements of mahogany (see Chapter 6) will encourage vigorous growth.
- *Creating conditions of lateral shade and overhead light.* Suitable conditions for encouraging strong vertical growth and recovery from attack can be achieved by planting mahogany in lines through natural forest, in mixtures with other tree species, or in pure stands at high densities. Thinning may be delayed in heavily attacked stands to allow trees to recover their form. Careful thinning of planting lines and mixed plantations is required to maintain overhead light and prevent suppression of mahogany trees.
- *Pruning* Provision must be made for pruning at regular intervals over the first 5 years of growth. Pruning should attempt to encourage vigorous growth of the leading shoot by removal of competing branches, rather than focusing on continuous removal of damaged areas.

11.2 Chemical control of *Hypsipyla* spp.

Newton *et al.* (1993a) review experience with different methods of chemical control, on which the following account is based.

Attempts to control *Hypsipyla* by spraying insecticides have repeatedly failed because of high rainfall in mahogany-growing areas and the fact that the larvae are inaccessible (Grijpma 1974, Allan *et al.* 1976, Wagner *et al.* 1991). For example, in Peru, trials with lead arsenate, DDT and Parathion applied at 2-weekly intervals were abandoned as a result of a low success rate and high costs (Doroujeanni 1963). Similarly in Venezuela, Ramirez Sanchez (1966) investigated control of *Hypsipyla* in small plantations of *C. odorata*, by spraying with DDT, metasystox, endrin, aldrin, parathion and combinations of these insecticides. DDT was found to be most effective, but repeated applications at every oviposition period (every 6 weeks) were required for effective control. This kind of application is clearly impractical on economic and ecological grounds (Wilkins *et al.* 1976).

The effectiveness of controlled release systemic insecticides was investigated in detail by Wilkins *et al.* (1976). Of a total of 28 different insecticides that were screened, 5 were found to be particularly effective in terms of pest control and lack of phytotoxicity. These 5 were tested in a series of field trials at CATIE, Costa Rica. The insecticides were applied in pellet form (consisting of polyamide resin) at planting. Carbofuran was found to be the most effective, giving complete control for 340 days at one of the sites tested. This was accompanied by lower mortality and higher growth rates of the treated trees (Wilkins *et al.* 1976). However, when a polymer preparation of carbofuran was tested in Trinidad, application of the insecticide was found to be completely ineffective in controlling shoot borer attack (Ramnarine 1992).

Controlled release insecticides may provide economic and ecological advantages when compared with conventional spraying (Allan *et al.* 1976). However, further research is required on the effects of soil and climatic conditions on the effectiveness of the treatments (Wilkins *et al.* 1976). The uptake and translocation of insecticide during dry periods may be ineffective (Grijpma 1974). The impact of systemic insecticide/polymer combinations on other insects, parasites and predators also needs investigation (Grijpma 1974). The main use of systemic insecticides may be as an interim measure, for protecting plants in the nursery (Wilkins *et al.* 1976).

Work currently being carried out at Luasong, Sabah, involves the use of insect pheromones (Matsumoto pers. comm.). The aim is to lure the males to traps, reducing the mating chance with females or to affect the location of females by the males, so that males are inactive. This method is considered good for large-scale control, since pheromones are easily distributed along enrichment lines. However, a similar approach was recently tested at CATIE in Costa Rica, without success; the approach depends on identifying a pheromone which is effective (Watt pers. comm.).

11.3 Biological control of *Hypsipyla* spp.

Experience with biological control of *Hypsipyla* has recently been reviewed by P. Baker in Newton *et al.* (1993a), from which the following account is taken.

H. robusta has been studied in the most detail; Rao (1969) and Rao and Bennett (1969) list more than 50 species of parasitoids and predators, including 17 braconids, 9 chalcids, 1 elasmid, 1 eulophid, 1 eurytomid, 13 ichneumonids, 2 trichogrammids, 2 tachinids, 1 nematode and 2 coleopterans. Nearly all of these were found in India from Uttar Pradesh, Punjab, West Bengal and Himachal Pradesh, and they are mostly larval and pupal parasitoids. Most were found in very low numbers (i.e. infestation rates of < 1%). A few were more common, including *Apanteles leptanus*, *Flavopimpla latiannulata* and *Tetrastichus spirabilis*.

Fewer natural enemies have been recorded for *H. grandella* in the New World; Rao and Bennett (1969) listed only 12, including 5 braconids, 2 ichneumonids, 2 trichogrammids, 2 tachinids and a mermethid. Since then a few more have been found (Grijpma 1972, 1973, de Santis 1972), but there are still considerably fewer known than for *H. robusta*, although this may simply be because less effort has been made to look for them.

It is clear that in both the New and Old World natural enemies do not control *Hypsipyla* populations sufficiently for plantation purposes. However, biological control may be possible with human intervention. A number of strategies have been suggested, including classical, augmentative and microbial biocontrol.

11.3.1 Classical biocontrol

Classical biocontrol involves the release of new species of natural enemies to a region in the hope that they will proliferate and control the pest without further human intervention. Over the last century this technique has been used successfully with a variety of insect pests, though there have also been many failures. A serious attempt was made in the 1960s and 1970s to introduce parasitoids from India to Trinidad and other parts of the Caribbean to control *Hypsipyla* (Cock 1985). As there were so few natural enemies in Trinidad, it was reasoned that one or more exotic species might be able exert control. *H. robusta* populations are not actually controlled by parasitoids in India, but it is quite possible that such biocontrol agents, in the absence of their own natural enemies (hyperparasitoids, predators and pathogens), can perform better in an exotic than in their native environment. It is also common for the natural host of the parasitoid to have evolved some immunity to attack, for example by a physiological response to the presence of a foreign egg in its haemocoel; there is a continuing debate amongst biocontrol workers about whether new host-parasitoid associations might be more effective than old ones (Hokkanen and Pimentel 1989).

In the case of Trinidad at least 6 species were imported from India, though appreciable numbers of only 4 species were released: *Tetrastichus spirabilis* Waterston (Eulophidae) (155 releases of a total of 103,050 over 10 years in 8 localities), *Trichogrammatoidea robusta* Nagaraja (118 releases, totalling 35,000 over 8 years in 9 places), *Phanerotoma* sp. (Braconidae) (142 releases of 42,540 over 8 years), *Anthrocephalus renalis* Waterston (Chalcididae) (26 releases of 3340). Appreciable numbers of these parasitoids were also released in Belize, Dominica, Grenada, St. Kitts, St. Lucia and St. Vincent (Cock 1985).

Only one, the egg parasitoid *T. robusta*, has been recovered consistently from the field in Trinidad and can therefore be said to have become established. Infestation rates of this parasitoid vary between 5 and 9%, but there is no evidence that it is having a significant effect. There are many possible reasons for the failure of this project, some of them related to costs and logistics. For instance, analysis of previous classical control projects suggests that the greater the number of a parasitoid species released, the greater the chance of successful establishment (Bierne 1975). Although the above release figures appear large, they were spread over many years and sites so that individual releases were small, no more then a few hundred at a time. Considering the great volume of a mahogany canopy and relative rareness of its host, the parasitoid could have great difficulty finding mates and oviposition sites. Cock (1985) implies that any future *Hypsipyla* project should involve substantial breeding programmes to increase the numbers to be released.

Other reasons for failure of the parasitoids to establish could be physiological, behavioural or semiochemical differences of the target host. Over the last few years it has become clear that the plant-host-parasitoid relationship is often very complex. Hérard *et al.* (1988) found that when a parasitoid emerges from its host it may become imprinted on specific chemicals (synomones) in the nearby frass of its host species and thence use that odour to locate a host in which to oviposit. From an unpublished report of introductions in the Caribbean for the years 1966-75 (Yaseen and Bennett 1975), it is apparent that *Hypsipyla* host material was often scarce and that alternative host species were used to culture the insects before release. It would seem possible therefore that the emerging insects could have been inadvertently conditioned to look for the wrong host. It may not be coincidental that the only natural enemy to become established was an egg parasitoid that presumably would not become imprinted on frass synomones.

There are still many parasitoids that could be introduced, but the costs of culturing the necessary *Hypsipyla* host material in order to enable the release of large numbers of enemies could be high.

11.3.2 Augmentative biocontrol

Augmentative control can be used where classical biological control is insufficient to drive pest populations below the desired economic threshold. This threshold is reached by an augmentative release of biological agents. A commonly cited example is the massive release of *Trichogramma* spp. which appears to work well under certain circumstances in the Cauca valley in Colombia (Holl *et al.* 1990).

In the case of *Hypsipyla* and mahogany there is a requirement for a very low insect attack rate. The economic threshold for this insect has never been calculated but will be extremely small, perhaps in the order of 10 attacks per hectare per year for out-planted saplings. The threshold must remain at a low level for about 5 years; i.e. until the unbranched stem has achieved a height such that subsequent damage will not significantly affect the market value. Few crops have more stringent requirements. It is hard to envisage that natural enemies could maintain such low levels constantly without augmentation. At present, production facilities for massive releases have yet to be developed, and it would be difficult to justify such an approach given the large and inaccessible areas over which such a scheme would have to operate.

11.3.3 Microbial control

Myers (1935) reported a *Cordyceps* sp. fungus from *H. grandella* in Trinidad. Rao (1969) recorded disease levels of 4-16% of *H. robusta* larvae sampled, but the cause was not identified. There have been various reports of susceptibility of *H. grandella* to pathogens; in the laboratory, larvae have been killed with *Bacillus thuringensis* (Hidalgo-Salvatierra and Palm 1972), *Beauvaria bassiana, B. tenella* (Berrios and Hidalgo-Salvatierra 1971) and *Metarhizum anisopliae* (Hidalgo-Salvatierra and Berrios 1972).

The general lack of reports of *Hypsipyla* disease in the field leads one to suspect that it is fairly uncommon. Given the low population density compared with many other insects and the protection conferred on the larvae by its concealment inside the mahogany stem or seed, this is perhaps not surprising. It is difficult to imagine a practical control strategy in plantations based on pathogen applications. The only possible use for pathogens that can be envisaged would be prophylactic sprays on nursery material.

11.4 Summary

- Of the many pests and diseases which affect mahogany plantations, *Hypsipyla* spp. (shoot borers) are the most significant. Two species are known to attack mahogany: *H. grandella* (Zeller), found in Central and South America and the Caribbean; and *H. robusta* (Moore) found throughout tropical Africa, Asia and Australasia. With the exception of certain oceanic islands (principally in the Pacific), *Hypsipyla* spp. appear to be present in all countries with suitable mahogany-growing conditions.
- Considerable efforts have been made to reduce shoot borer damage by silvicultural means. Silvicultural control may be achieved through a number of mechanisms: preventing the *Hypsipyla* sp. from locating the mahogany plants; reducing the susceptibility of mahogany plants to shoot borer attack; increasing the tolerance of mahogany plants to shoot borer attack; or encouraging natural enemies of the *Hypsipyla* sp.
- To take advantage of these mechanisms, silvicultural techniques include selection of resistant provenances, selection of suitable sites and careful design of the plantation (in terms of layout and species composition). After plantation establishment, managers may use various weeding, fertilising, pruning and thinning techniques to help limit attack and aid recovery.
- Particular emphasis should be placed on silvicultural techniques which facilitate recovery after attack, through promoting vigorous vertical growth and restricting the development of lateral branches. This can be achieved by careful site matching, techniques which create conditions of lateral shade and overhead light, and pruning.
- It is unlikely that any single silvicultural technique will successfully control shoot borer damage and a combination of locally suitable techniques should be adopted. Regional variations in both shoot borer and mahogany populations and the possible effects of site on pest-host interactions mean that techniques which have been successfully used on one site may be ineffective on another.
- Chemical control of *Hypsipyla* spp. in plantations is not generally recommended on economic and environmental grounds. However, systemic insecticides may be useful in preventing damage in the nursery.
- Research on biological control of *Hypsipyla* has failed to identify an effective agent. Successful implementation of this approach will require further research and will be limited by the high cost involved and concerns over releases of exotic species. Silvicultural techniques which encourage natural biological control may be more suitable.

Chapter Twelve

Protection

Interest in the shoot borer has diverted attention from other pests and diseases known to damage mahogany plantations. Ambrosia beetles and termites have caused particular problems in Fiji and may pose a threat to large plantation areas in other countries. Mahogany is often grown in parts of the tropics affected by hurricanes and, to a lesser extent, fire. In this chapter, measures for protecting plantations from associated damage and treatment of affected stands are considered.

12.1 Control of pests and diseases

12.1.1 Ambrosia beetles

Ambrosia beetles are representatives of a number of Coleoptera families. The beetles that attack mahogany are from two closely related families, the Platypodidae and the Scolytidae (Roberts 1973). The adult beetles tunnel into wood creating a 'pin-hole' which is visible in sawn timber. An ambrosial fungus, carried in special mycangia, may be introduced (Speight and Wainhouse 1988). The beetle feeds off the fungus, which in turn feeds off the tree, staining the timber around the pin-hole galleries a dark shade (Oliver 1992). In living mahogany trees, the beetles are usually killed at an early stage by gum exudation produced (SRD 1992). The ambrosial fungus does not have a chance to grow, but the pin-holes reduce the grade of sawn timber. A number of species of ambrosia beetle have caused pin-hole damage to mahogany trees (see Table 12.1).

Fijian mahogany, which has suffered the most serious incidence of attack, was found to be most susceptible from years 6 to 8 after planting (see Box 12.1; Oliver 1992). However, in other countries, beetles have been observed to attack younger trees (see Table 12.1). A number of silvicultural techniques have been suggested to limit ambrosia beetle attack (Soubieux 1983):
- regular cleaning of the plantation;
- rapid removal of thinnings and fellings from site;
- avoidance of sites prone to drought;
- use of mixtures;
- removal of attacked trees (two cuttings spaced 1 month apart gave good results).

Other measures have been recommended for limitation of damage in Fiji (see Box 12.1).

Table 12.1. Species of ambrosia beetle reported to damage mahogany

Species (Family) [Country]	Nature of attack	Source
Crossotarsus externe-dentatus (Platypodidae) [Fiji]	See Box 12.1 for details	SRD (1992)
Hexacolus guyanaensis (Scolytidae) [Trinidad]	The beetle attacked pole stage mahogany up to 7 yrs old, breeding in dead wood on the forest floor	Pawsey (1970)
Hypothenmus eruditus (Scolytidae) [Malaysia]	The beetle attacks young seedlings	Ata & Ibrahim (1984)
Platypus gerstaeckeri (Platypodidae) [Fiji]	See Box 12.1 for details	SRD (1992)
Xyleborus morstattii Scolytidae) [Indonesia]	The beetle bores into living twigs and branches	Bramasto & Harahap (1989)
Xylosandrus compactus [Puerto Rico]	The beetle attacked stems of mahogany saplings <3 years old and twigs <25 mm in diameter. In a young plantation, 70% of trees were infested, but healthy trees were not badly affected. The original habitat of the beetle was probably dead wood from trees felled to clear enrichment lines	Bauer (1987b)
Species unknown [Guadeloupe]	Ambrosia beetles caused significant damage to mahogany plantations. Unhealthy or dead trees provided breeding grounds. Healthy trees were susceptible in periods of drought	Soubieux (1983)

Box 12.1. Ambrosia beetle damage to plantations in Fiji

By 1971 a considerable area of mahogany plantation had already been established in Fiji. Evidence that the ambrosia beetle was present and causing damage therefore caused considerable concern and the planting programme was halted in 1972 (Roberts 1977, JICA 1982). A survey found the beetles present in all main plantation areas, but with wide variations in the degree of infestation within individual plantations. Two species were identified: *Crossotarsus externe-dentatus*, which is widely distributed throughout the tropics, and *Platypus gerstaeckeri*, endemic to Fiji (Roberts 1977). Neither beetle is species-selective, although *P. gerstaeckeri* was found to be size-selective, rarely attacking main stems below 15 cm in diameter (SRD 1992). Since most of the plantations were relatively young in 1972 and stem diameters small, attack by *P. gerstaeckeri* was less widespread than *C. externe-dentatus*. Unfortunately, *C. externe-dentatus* is the more aggressive of the two species and therefore more likely to cause damage.

Around 20% of trees were affected (Oliver 1992). Trees located at the edges of compartments were most commonly attacked (Roberts 1977). Often roadworks had occurred nearby but trees were never obviously damaged by these activities. Inside plantations, attack was usually related to some forest operation such as thinning, pruning, cleaning or the removal of sample trees for experiment, but untreated stands also suffered light attack. Sites that had been extensively damaged by a cyclone were badly affected. Attack began from 1 to 3 weeks after the beginning of the disturbance and, where the vegetation matrix was drastically changed, occured for up to 5 months afterwards. The number of trees attacked and the abundance of ambrosia beetles, *P. gerstaeckeri* in particular, appeared to be directly related to the intensity of cleaning. In intensively cleaned stands, indigenous trees that were left in amongst the mahogany were not attacked. Attack was also found to be related to site quality: serious attack was found in areas of poor drainage and on poor soils (Roberts 1973).

It appears that attack is related to some form of stress on the mahogany trees (Roberts 1977). In intensively cleaned stands, it was suggested that the attack was a result of water stress. Sudden exposure of crowns to direct sunlight after intensive cleaning increased evapotranspiration rates and causes a temporary water deficit, which may have affected the trees' defence mechanisms (Roberts 1973). Damage to the roots of trees on the edge of plantations may have affected water uptake and thus produced similar results. Roberts (1977) suggests that the absence of attack on indigenous trees indicates that mahogany is poorly adapted to Fijian sites. However, in other respects, the trees are generally healthy and growing well.

Box 12.1. (continued)

Given that the ambrosia beetles are unable to develop suitable galleries in living mahogany trees from which to breed (Singh pers. comm.), there must be other breeding grounds in the vicinity. The presence of large beetle populations is probably related to the techniques used to establish mahogany plantations (Roberts 1977). Residual natural forest canopy trees were poison-girdled before new plantations were established. The poisoning was often only partially effective, with many trees dying slowly over a period of years. Under such conditions, ambrosia populations can be sustained for 4 years or more. When breeding grounds in decaying trees were finally depleted, the beetles emerged to attack the planted mahogany. Alternative sources of breeding material are thinnings and prunings left on the forest floor and woody debris produced by cyclones.

In the light of the Fijian experience, a number of rules were suggested for protecting new plantations from ambrosia beetle attack (Roberts 1973, 1977):
- where logging has been carried out in natural forest, allow 5 years before line planting to allow beetle populations to subside
- poison as early as possible, ideally just before planting and not later than 5 years after planting
- limit poisoning to trees which overshade the mahogany
- fell trees which do not respond to poisoning (e.g. *Myristica castaneifolia*)
- after 6 months repoison trees which are not yet dead
- clean lines 3.6 m wide, reducing poles and shrubs to a height of 30 cm
- plant mahogany seedlings in wet weather
- continue line cleaning for 4 years after planting, leaving a ground cover of vegetation but allowing the young mahogany to dominate
- after 4 years, restrict cleaning to climber cutting, ensuring that vegetation cover around the base of the tree is maintained
- at the time of first and second thinnings of mahogany, cut stems as close to the ground as possible; avoid thinning adjacent trees and damaging vegetation at the base of remaining trees
- prune with care
- remove all thinning, pruning and hurricane debris from the stands
- reduce the number of silvicultural operations to a minimum to avoid disturbance

Although there are a number of natural predators which will eat ambrosia beetles, these have not been effective in reducing beetle populations in plantations (Roberts 1977).

The ambrosia beetle scare led to a cessation in mahogany planting for 2 years. However, planting has continued ever since and modification of silvicultural techniques, including a moratorium on thinning, has reduced the extent of the problem to an acceptable level (Singh pers. comm.). Damage to trees that suffered the beetle outbreak in the 1970s is mostly confined to a 10-15 cm timber core. Although the value of sawn timber has been reduced, pin-holes are sparse enough not greatly to reduce the value. The value of peeler timber is largely unaffected, since the central core cannot be utilised by the peelers. Termites now appear to pose the greatest threat to the Fijian mahogany plantations (see Box 12.2).

12.1.2 Termites

The extent of termite damage to mahogany plantations is probably underestimated. Recent work in Fiji (see Box 12.2) indicates that the insect is responsible for attacks on a surprisingly large scale. Species of termite have also been observed to cause damage in: Sri Lanka; the Solomon Islands, where a species of *Coptotermes* feeds on living wood (Oliver 1992); and Belize, where 11.5% of managed pole stage regeneration in natural mahogany forests was attacked (Stevenson 1940). Damage to mahogany stems may not become apparent until termite galleries are well developed (see Plate 12.1). Secondary infection of galleries by fungi can obscure the fact that termites were the original cause of the damage. It appears that termite damage can be limited by selection of appropriate planting sites; effective biological control of termite damage to mahogany may also be possible in the near future (see Box 12.2).

Plate 12.1. Symptoms of well-advanced termite damage. Nukurua plantation, Fiji.

Box 12.2. Termite damage to plantations in Fiji

Termites have replaced the ambrosia beetle as the main problem for mahogany forest managers in Fiji (Oliver 1992, Kamath pers. comm.). In a study in 1993, termite incidence in plantation trees was found to range from 6 to 8.5%, based on a sample size of 6081 trees (Kamath et al. 1993). Incidence in extracted logs varied from 5.7 to 9.5%, based on a sample size of 1491 logs. Loss in economic value of the timber was found to be 8%. Local loggers will cross cut until they find a clean section, so all affected timber is wasted, even if only the core of the log is affected (Kamath pers. comm.). It is now estimated that numbers of affected trees may be vary from 5 to 22%. Termites therefore pose a major threat to present and future mahogany plantations (Kamath et al. 1995).

Although there are 9 species of termite reported in Fiji, only 3 (*Neotermes samoanus*, *N. papua* and an unidentified species) have been observed to attack living mahogany (Kamath et al. 1993, Kamath et al. 1995). Other species, notably *Procryptotermes* spp., attack dry mahogany timber, but will not affect trees in plantations (Kamath et al. 1995).

Termites affect trees of all ages, including natural regeneration, regardless of tree health (Kamath et al. 1993). Forest managers are not aware of the attacks because they occur below ground (Oliver 1992), and attacked trees may initially have no outward symptoms. However, as the attack progresses, gentle to heavy swellings appear on the bole, giving it a wavy profile. Deep cracks with oozing gum may be visible on the bark. The tree may appear to grow normally, even when the attack is severe.

The most important breeding grounds for termites are tree stumps. Infestation of surrounding trees occurs by the establishment of satellite colonies linked to the main nest by underground galleries (Kamath et al. 1993). Attack is therefore localised. The method of infection explains the lack of observed correlation between infestation and site type.

Box 12.2. (continued)

Neotermes invades the root of mahogany trees and moves up the tree inside the stem (Kamath *et al.* 1993). The termites excavate a 'pipe' up through the middle of the bole. This may be associated with secondary galleries leading horizontally into the heartwood and sapwood at various intervals (Kamath pers. comm.). The frass-filled galleries follow the annual rings (Kamath *et al.* 1993). The cracks which are observed in the bark serve as air vents for termite galleries and indicate that they are very close to the surface of the trunk. The galleries may extend 10 m up the inside of a tree. Ultimately, the root collar, large roots, buttresses and the bole may be hollowed out.

Termite infestation is occasionally associated with heart-rot fungi (*Armillaria mellea* or *Phellinus noxius*) (Kamath *et al.* 1993). Like the termite, the fungi spreads underground from infested roots or logs. Initially it was thought that fungal and termite attack were linked, and that fungal attack preceded termite attack (Kamath pers. comm.). However, very few of the trees studied so far contain evidence that termite and fungi have been present and no sporocarps of either species of fungi have been found.

No termite damage has been observed where mahogany plantations have been established in areas without stumps (Kamath *et al.* 1993). However, in Fiji mahogany plantations are established on logged natural forest sites, where the presence of stumps is unavoidable (Kamath *et al.* 1995). Furthermore, today's plantations are nearing the end of their rotation and there is a real concern that infested mahogany stumps could become the focal points of infestation during future replanting programmes (Kamath *et al.* 1993).

A limited number of preventative measures are available. Stumps may be uprooted, but this option is far too expensive to be practical. Stumps may be chemically treated, but chemical control is also expensive since treatments must be continuously and regularly applied. The Forest Department in Fiji is discouraging the use of chemical applications for environmental reasons and would not favour such regular usage. The other option is biological control. Trials that have been carried out so far indicate that this method is extremely successful.

A pathogenic fungus, *Metarhizum anisopliae* (strain Fl610) has been used (Kamath pers. comm.). The fungus may be mixed with talcum powder (ratio 1:1) to maintain dryness and viability of the spores. Infected stumps or trees are located and holes are drilled into the stem until the main termite gallery is located (this is seen from the cores which are extracted). On average it takes 3 attempts to locate the main gallery. The spores of the fungi are then pumped into the hole and the hole is sealed. The spores come into contact with the termites and are carried throughout the colony. The spores are able to grow into the body of the termite and the mycelia multiply, causing death. After 2 months (or sooner in many cases) the whole colony can be wiped out. Trials carried out since 1995 indicate that the method of control is highly effective. The main drawback appears to be the repeated coring required to locate the main gallery.

However, no large-scale plantation application has yet begun (Kamath pers. comm.). More information is required on the longevity of the fungus once injected into the tree and whether or not the termites are able to re-invade treated trees. There are other biological agents, untried in Fiji, which could also be tested. A nematode (*Heterorhabditis* sp.) has been used successfully for controlling termites in Australia, Sri Lanka (in tea plantations) and Tuvalu (in coconut plantations). The worm enters the body of the termite and releases a toxic bacterium, *Xenorhabdus* and/or *Photorhabdus* spp. The worm is very effective in spreading though galleries, but requires a highly moist environment.

12.1.3 Other pests and diseases

A number of other pests (see Table 12.2a) and diseases (see Table 12.2b) have been reported on mahogany trees. In general it appears that incidences of attack have not been serious, since methods for control are not widely discussed in mahogany-related literature. Most of the pests and diseases described are common in plantations of other species and their control is described in more general literature.

Table 12.2a. Pests that are known to damage mahogany plantations

Pest [country] *Scientific name*	Nature of damage and suggested control (if any)	Source
Leaf and stem		
Agouti [Costa Rica]	Uproots seedlings while foraging for seeds	Gerhardt (1994)
Boar [Indonesia, India]	Uproots young saplings and strips bark off trees	Wind Hzn (1926), Troup (1932)
Coreid bug [Solomon Islands] *Ambleypelta cocophaga*	Causes die back of terminal bud. **Control:** Shade terminal buds with natural vegetation	Oliver (1992)
Cow [Vanuatu]	Chews bark off young trees	Leslie (1994)
Deer [India, Indonesia]	Browses young saplings; strips bark off trees	Wind Hzn (1926), Troup (1932)
Defoliator [Belize, Honduras] *Egchiretes nominus*	Attacks trees during the rainy season, esp. trees under shelterwood where humidity is high. **Control:** Remove shelter-wood; promote vigorous growth	Stevenson (1944b), Chable (1967)
Goat [India]	Browses young saplings	Troup (1932)
Leaf cutter ant [Honduras]	Strips leaves on recently planted trees	Chable (1967)
Mistletoe [India] *Dendropthoe falcata*	Capable of serious damage to the tree	Gibson (1975)
Squirrel [Sri Lanka, Malaysia, Indonesia]	Strips bark off young and sometimes old trees	Wind Hzn (1926), Streets (1962), Muttiah (1965)
Weevil [Puerto Rico] *Diaprepes abbreviatus*	The adult feeds on fresh leaves, but does not cause complete defoliation. More serious damage (even mortality) is caused by the larva feeding on the root stalk below ground level	Bauer (1987a)
Weevil [Malaysia] *Dysercus longiclaris*	Ring barks and kills young mahogany trees	Streets (1962), Ata & Ibrahim (1984)
Seed		
Parrot [Mexico, Belize]	Eats seeds	Santiago et al. (1992), Stevenson (1944b)
Rat [Fiji]	Eats seeds	Evo (pers. comm.)

Table 12.2b. Diseases that are known to damage mahogany plantations

Disease [country]	Mode of transmission and nature of damage	Suggested control	Source
Root			
Armillaria mellea [Fiji, Indonesia]	Spreads from stumps of poisoned or logged natural species, through root contact. The disease can spread along and sometimes between mahogany planting lines, particularly when trees mature and roots come into contact. It produces a slight swelling in the butt region, accompanied by resin exudation and occasionally bark cracking. A thick, white mycelial mat develops between the bark and the wood of disease parts. Causes rot of the bark and wood from the roots and root collar. Attack by termites and ants follows. May kill older trees	It may be sensible to delay planting for 2-3 years after natural vegetation has been removed, to give stumps more time to decay. Wide spacing between trees and lines will help to reduce spread of the disease. Dig a trench around infected trees in young plantations	Gibson (1975), Singh & Bola (1981), Bramasto & Harahap (1989), SRD (1994)
Phellinius fatuosus [India]	Enters the tree through wounds, causing root rot and increasing susceptibility to windthrow. Large honeycomb pockets are filled with white fibres; later, when empty, the pockets become dark brown	See below	Gibson (1975)
Phellinius noxius [Sri Lanka, Fiji, Indonesia, Vanuatu]	Spreads from stumps of poisoned or logged natural species, through root contact. The first symptoms are the formation of a dark encrustation on the roots and bark of the lower stem which has a thick, white margin. Small, hard, dark-brown brackets with a grey poroid undersurface are sometimes produced. Within the affected area, cracking and gummosis may occur. In trees under 10 years old, the disease may girdle the stem causing sudden death. In older trees, adjacent cambium may spread over the infection, producing callusing and longitudinal cankers. Branches in the crown on the infected side of the tree may die, increasing susceptibility to windthrow. The fungus may kill mature trees. In Vanuatu the worst outbreaks of the fungus are associated with close spacing (3 x 3 m)	A trench (0.7 m deep) between infected and healthy trees limits spread in young plantations. It may be sensible to delay planting for 2-3 years after natural vegetation has been removed, to give stumps more time to decay. Plant lines at a wider spacing or interplant mahogany with resistant trees	Busby (1967), Gibson (1975), Singh *et al.* (1980), Oliver (1992), Leslie (1994), SRD (1994), Evo (pers. comm.)
Pratylenchus coffeae [Barbados]	Attacks the root cortex and stem base, invading the cambium and underlying phloem. The trunk becomes gnarled and contorted, with dark staining and scaling of bark up to shoulder height. It is not clear whether the disease affects the health of the tree directly, but affected regions are often invaded by secondary micro-organisms	nda	Gibson (1975)

Table 12.2b (continued). Diseases that are known to damage mahogany plantations

Disease [country]	Mode of transmission and nature of damage	Suggested control	Source
Root			
Pythium splendens [St. Lucia]	Spreads from root to root causing rot which spreads up through the heart of the tree. Symptoms are yellowing and shedding of leaves and lifting of bark at the tree base	Selectively fell diseased stems and/or reduce stocking density	John (pers. comm.)
Rhizoctonia sp. [St. Lucia]	Spreads from root to root causing rot which spreads up through the heart of the tree. Symptoms are yellowing and shedding of leaves and lifting of bark at the tree base	Selectively fell diseased stems and/or reduce stocking density	John (pers. comm.)
Unknown fungus [Indonesia]	Spreads from stumps of *Ficus elastica* to mahogany	Uproot *Ficus* stumps and cultivate soil for 2-3 years before planting	Hart (1925)
Leaf and stem			
Botryodiplodia theobromae [Philippines]	Causes stem rot	nda	Soerianegara & Lemmens (1994)
Cercospora subsessilis [India, Peru, Puerto Rico]	Causes leaf spots in the form of zonate, pale brown to grey lesions of 2-5 mm in diameter	nda	Gibson (1975)
Glomerella cingulata [India]	Causes leaf spots	nda	Gibson (1975)
Micosphaerella spp. [Cuba, Venezuela]	Causes leaf infection	nda	Gibson (1975)
Thread blight [Indonesia]	Grows on underside of branches on older trees in the form of white mycelial threads, attacking leaf petioles and causing leaves to wither	nda	Gibson (1975)
Viral canker [Trinidad]	Causes swelling of main stem and rupturing of bark. May girdle and kill trees less than 30 cm dbh. The canker is usually located less than 4 m from ground level and is up to 90 cm long by 60 cm wide on old trees (though does little damage to surrounding timber). Infects natural regen.	nda	Streets (1962), Pawsey (1970), Gibson (1975)

nda = no details available

12.2 Protection from hurricanes

Hurricanes (known as cyclones in the Pacific and typhoons in South-East Asia) are a feature of many mahogany-growing countries, including the Philippines, Vanuatu, Fiji, Western Samoa, Belize, Mexico, Guatemala and the Caribbean islands (Lindo 1968, Neil 1986, Oliver 1992, Snook 1993, Soerianegara and Lemmens 1994, Singh pers. comm.). In these countries, mahogany is often preferred to other plantation species because of its ability to survive strong winds with relatively little damage to the crown. Possible reasons for this resistance have already been discussed (see Section 2.5.4).

12.2.1 Comparative resistance to hurricane damage

There is much evidence to indicate that mahogany is more resistant than most other tree species to hurricanes. In Belize, large mahogany trees in natural forest were found to be more resistant to hurricanes than other species of similar size (Lindo 1968). After hurricane Hattie, which struck Belize in 1961, young mahogany trees in plantations survived extremely well when compared with surrounding natural regrowth, which was effectively 'weeded out' (Frith 1962, Langley 1963). Mahogany trees flushed a few weeks after the hurricane and small broken trees coppiced well (Frith 1962). The most serious damage to young mahogany trees was caused by other trees falling on top of them.

Similar resistance has been noted in mahogany plantations of South-East Asia and the Pacific. Only winds of exceptionally high velocity caused damage to mahogany in the Philippines, although saplings were more susceptible (Ponce 1933). After two cyclones (Eric and Nigel) which struck Vanuatu in 1985, it was concluded that mahogany was very windfirm and showed only a moderate tendency to lose foliage and snap (Neil 1986). Cyclone Ofa (1989) caused around 40% destruction to plantations of fast-growing species in Western Samoa but only 14% to mahogany (Oliver 1992). The Western Samoan Forest Department have since switched to slow-growing species, principally mahogany. Cyclone Kina struck Fiji in 1993 causing appalling damage to pine plantations (SRD 1993). In contrast, less than 15% of mahogany trees were found to have damage worse than 'very slight' (i.e. a few small branches broken off and leaves stripped).

Despite the ability of mahogany trees to survive hurricanes better than most other species, damage may still be substantial. In Belize, hurricanes were found to reduce growth rates of mahogany in natural forest considerably. Lamb (1947) noted that after the 1942 hurricane, mahogany trees in Freshwater Creek took 3-4 years to recover their pre-hurricane diameter growth increment: growth from 1942 to 1944 averaged only 0.1 cm yr^{-1}, rising to 0.3 cm yr^{-1} in 1945 and 1.0 cm yr^{-1} in 1946.

12.2.2 Limitation of hurricane damage

Foresters in Fiji now believe that root pruning adversely affects the long-term development of root systems, by reducing the development of a strong tap root and thus the ability of trees to withstand cyclones (Singh pers. comm.). Root pruning in the nursery is no longer practised and 50% of new plantations are being established by direct seeding in order to minimise root disturbance.

Neil and Barrance (1987) recommended a number of measures to ensure that damage caused by cyclones is minimised, including:
- careful site selection, avoiding shallow soils and soils prone to waterlogging;
- retention of residual natural forest around plantations as windbreaks;
- the use of short rotations, although fast-growing mahogany trees are more susceptible to wind damage (Frith 1962, Lindo 1968).

Mayhew et al. (in press) have recommended the use of an uneven-aged group selection system for mahogany plantations in hurricane zones. Although the older trees in the stand are susceptible to windthrow, the younger trees are given some protection and have a higher chance of survival, reducing the risk of plantation failure. A canopy gap created by a hurricane can be treated like any other felling coup and thus easily incorporated into the silvicultural system. Once the windthrown trees have been cleared, natural regeneration can be encouraged.

However, no amount of advance planning will protect against the worst hurricanes and it is useful to have contingency plans should a disaster strike. In Western Samoa, saplings which have blown over are propped up with stakes (Oliver 1992). This technique must be implemented quickly, before the stem has a chance to bend upwards. In some countries, it may be feasible to secure markets for small diameter mahogany timber in advance (Neil and Barrance 1987). Markets can be supplied with thinnings between hurricanes. For older plantations, salvage logging may be necessary, although remedial action is not always economically viable (Francis pers. comm., Seroma pers. comm.).

12.3 Protection from fire

12.3.1 Comparative resistance to fire damage

The ability of mature mahogany trees to survive natural fires has been cited as one of the reasons for the presence of the species in natural forests of Central America (Snook 1993). Yet mahogany has no distinctive fire-adapted traits apart from the thick bark on old trees (Lamb 1966, Lamprecht 1989). In plantations, mahogany is considered to be a fire-sensitive species (Burger 1924, Lamprecht 1989). Frequent grass fires in Piña Blanca plantation, Tuguegarao, Philippines, caused mortality of many saplings in the period immediately after establishment (Patricio pers. comm.). Mature trees in Sri Lanka were also adversely affected by fires. Sandom and Thayaparan (1994) found that fire damages surface roots and destroys the bark of the lower stem, greatly increasing the susceptibility of the tree to decay. However, fire did not necessarily cause mortality.

There are two points which are critical to the interpretation of these observations. Firstly, variations in the intensity of the fire associated with fire damage are rarely recorded. Mahogany may not be able to survive high intensity fires, but there is some evidence that older trees are tolerant to low-intensity ground fires (Streets 1962, Patricio pers. comm.). Secondly, differences in opinion over the sensitivity of mahogany to fire depend on the perspective of the observer. The ability of fire-affected trees to survive and reproduce for many years might indicate fire-tolerance to an ecologist; but to a forest manager, whose prime concern is timber yield, damage to and subsequent decay of timber would indicate high sensitivity to fire.

12.3.2 Limitation of fire damage

In general, fire is not prevalent in mahogany-growing countries because of the climatic conditions (high annual rainfall and/or short dry seasons) required by the species. Mahogany plantations in regions with a pronounced dry season are most at risk from fire, particularly when surrounded by farms and villages (as in Sri Lanka). In such cases, use of standard preventative measures, such as firebreaks, may be required.

After a plantation has been burnt, it may be advisable to remove damaged and unhealthy trees quickly before the timber decays and the site becomes a breeding ground for pests and diseases. In cases where the intention is to use natural regeneration to establish the next rotation, foresters should retain seed trees and refrain from salvage logging until natural regeneration has been established (Sandom and Thayaparan 1995). Experience from natural forests indicates that the conditions created after a fire should favour natural regeneration of mahogany (see Section 2.5.4).

Mahogany may be used to protect other, more susceptible species from fire. Bramasto and Harahap (1989) note that the leaves of mahogany are difficult to burn and that the tree can be used as a firebreak or in areas susceptible to burning. Gonggrijp (1940) reports the use of mahogany as a living firebreak in Indonesia to protect pine plantations. However, Burger (1924) stresses that mahogany is not suitable as a firebreak species.

12.4 Summary

- Ambrosia beetles are known to cause damage to young mahogany plantations in some countries. To reduce risk of attack, dead wood should be removed from the planting site. Where risks of attack are high, silvicultural operations (particularly thinning) within the plantation should be minimised.
- The damage caused by termites to plantations may be significant. Affected trees are susceptible to windthrow or windsnap. If possible, pre-existing stumps on the planting site should be removed or chemically treated. Biological control, through the introduction of a pathogenic fungus into termite colonies, appears to be effective.
- Mahogany is widely planted in regions affected by hurricanes, and is one of the most resistant plantation species to hurricane damage. Silvicultural techniques, which may enhance the natural resistance of mahogany to wind damage, include: direct seeding, to encourage development of a strong tap root; careful site selection, to encourage deep rooting; and adoption of uneven-aged silvicultural systems, to reduce risk of total destruction.
- Mahogany plantations in areas with a pronounced dry season may be at risk from fire. Mahogany is considered to be susceptible to fire; although trees may survive, timber subjacent to damaged bark may be subject to fungal attack. In the event of fire, it may be wise to carry out a salvage felling before timber quality deteriorates further.

Chapter Thirteen
Silvicultural Systems

The silvicultural techniques used for establishment and early tending of mahogany plantations have been well documented (see Chapters 7 and 8). However, the application of these techniques, termed 'reproductive methods' by Smith (1986), does not amount to the adoption of a silvicultural system. A silvicultural system is defined as 'the process by which the crops constituting the forest are tended, removed and replaced by new crops, resulting in the production of woods of a distinctive form' (Troup 1928). A silvicultural system implies the application of a planned programme of silvicultural treatment (Smith 1986) and addresses the establishment of second and subsequent rotations. The definition and adoption of a suitable silvicultural system is therefore an important step towards sustainable forest management, although the latter concept has broader, less production-orientated objectives.

Interest in reproductive methods rather than long-term silviculture is to be expected. The key stage in the life cycle of a mahogany plantation is during the period of early growth, when shoot borer attack may have a profound effect on the form of the trees and thus the value of the plantation. Furthermore, the existence of mature mahogany plantations is a relatively recent phenomenon and foresters in the past have had little need to consider the use of silvicultural systems. However, there are now a number of countries where mahogany plantations are nearing or have already reached maturity, including Martinique, Guadeloupe, Sri Lanka, Philippines, Indonesia and Fiji (Mayhew et al. in press). Careful consideration of long-term plantation management, including the establishment of the next rotation, is required if forest managers are to make best use of the opportunities presented by maturing plantations.

In natural mahogany stands, where mature trees are already present, foresters interested in ensuring the long-term productivity of the forest are immediately presented with the opportunity to use a range of silvicultural techniques which can encourage natural regeneration and the development of a future timber resource. Observations of the response of mahogany to silvicultural treatments in its natural habitat (for example, the response of natural regeneration to canopy opening) may be useful in helping foresters to decide on a suitable silvicultural system for plantations. Management experiences in natural forest, in particular the techniques used to encourage natural regeneration, are therefore discussed (see Section 13.1). Subsequently, the silvicultural systems which have been proposed for mahogany plantations are illustrated with three case studies (see Section 13.2).

13.1 Management of mahogany in natural forests

Examples of management of mahogany in natural forests are categorised as commercial scale management, which has been carried out profitably over large areas (see Section 13.1.1) or as research scale management, which has been carried out on an experimental basis over small areas (see Section 13.1.2).

13.1.1 Commercial scale management

The main obstacle to the development of silvicultural systems for the management of large natural forest concessions is the lack of an economic incentive for positive silvicultural interventions. In an interesting, but largely theoretical study, Howard *et al.* (1996) compared financial returns and environmental impacts of four alternative silvicultural prescriptions to a sample area in the Chimanes Forest of Bolivia. Two of the four prescriptions represented variations of current cutting practies, involving the higly selective removal of all commercial stems of mahogany and other valuable species on a 5 and 10 year cutting cycle The other two prescriptions (untested) were designed to retain mahogany seed sources and to concentrate production over a smaller area, in carefully controlled conversion and maintenance phases; more low-value species were cut, but improved conditions for natural regeneration of mahogany were created. Variations of current cutting practices were found to be much more profitable, with a net present value 3.5-7 times higher than the more sustainable alternatives. It is not surprising, therefore, that 'management' of mahogany in natural forest is typically limited to the imposition of simple cutting restrictions.

Cutting diameter restrictions have been in operation in natural forest in Belize since May 1920, when Crown Land Rules prohibited the cutting of mahogany 'squaring' less than 38 cm. The figure was later converted to a minimum cutting diameter of 73 cm (Stevenson 1944b). Problems arose at the outset with failure to enforce the rules (Oliphant 1925) and repeated illegal felling was reported by subsequent conservators (e.g. Kinloch 1938). By the 1940s it was noted that the reproductive capacity of mahogany in the northern part of Belize was reduced to its 'lowest limit' (Stevenson 1944b). The salvage felling that was sanctioned after hurricane Hattie in 1961 led to the loss of most remaining merchantable stems (Bird pers. comm.). As a result, natural forest in Belize is now largely depleted of merchantable mahogany down to a level significantly below that of the original minimum cutting diameter. For a more comprehensive historical account of the logging of mahogany forest in Belize, readers are referred to Weaver and Sabido (1997) or Chaloner and Fleming (1850).

Much of the natural mahogany forest in Bolivia has only recently been opened up to exploitation and mahogany stocks are still high. A recent management plan for the logging of the Chimanes Forest in Bolivia specified a minimum cutting diameter for mahogany trees of 80 cm diameter above buttress (Gullison 1995). In fact, trees with a diameter as low as 40 cm have been cut (Gullison in press). An additional constraint that 10% of commercial sized trees should be left as seed trees was also specified. However, where logging has taken place, the rules have been ignored. For example, of the 65 mahogany trees over 80 cm in Gullison's study area, only one was designated a seed tree. The long-term prospects for natural mahogany regeneration are bleak.

In Honduras, a management plan for natural mahogany forests along the Atlantic coast has been produced by the Proyecto Desarollo del Bosque Latifoliado (Meza pers.

comm.). Some of the forests, which are managed by 'sociedades collectivas', contain a high percentage of mahogany. The best examples are found in inaccessible regions. For example, at Las Mangas, Departamento de Colón, which consists mostly (around 92%) of primary rainforest, there are 5-6 large stems of mahogany or around 30 m^3 per hectare. The minimum cutting diameter for mahogany has been set at 50 cm. The production area (515 ha) is to be divided into 30 compartments, each with a felling cycle of 30 years. A total of 1950 m^3 yr^{-1} (all merchantable species) will be extracted at each cut. If there is insufficient regeneration (defined as less than 167 stems ha^{-1}), there are plans to plant mahogany in clearings, spaced at 6 x 6 m (leaving the shaded 6 m wide band at the edge of the clearing unplanted). Because of the nature of the sites (thin, loamy soils overlying rocky ground, steep slopes and many rivers) protection is an important aspect of management: of the total managed forest area at Las Mangas (721 ha), nearly 30% has been set aside for protection. The impact of logging operations has yet to be assessed.

Management of the mahogany-bearing forests in Quintana Roo, Mexico, is carried out through a community forestry initiative known as the Plan Piloto Forestal (PPF) (Snook in press). PPF organises forest 'ejidos', legally defined areas which are owned and run by local coummunities. The forests are currently being inventoried in detail, in order to establish merchantable volumes and to assess the response of residual forest to logging (Gutierres and Cabral pers. comm.). In Las Caobas ejido, the stocking density of merchantable mahogany trees averages 0.6 stems ha^{-1}, although it should be noted that there are 6 other marketable species and 8 more with potential markets (Hassan pers. comm.). However, management of mahogany is still based on a minimum cutting diameter of 55-60 cm and a cutting cycle of 25 years, a legacy of a federal body which held a large Quintana Roo concession from 1957 to 1983 (Snook in press).

Despite certification of ejido timber, there appears to be doubt as to whether the proposed management regime is sustainable. Some (e.g. Snook 1993, in press, Flachsenberg 1994) believe that specification of a minimum diameter and a 25 year cutting cycle is not sufficient to guarantee natural regeneration of mahogany. The main criticisms are that the gaps in the canopy are too small and that there has not been enough disturbance at ground level to create suitable conditions for germination and growth of mahogany seedlings. Attempts have been made to address these problems. Natural regeneration is complemented by planting on skid trails and log-landings and there are now plans to clear vegetation around trees designated for felling in an attempt to determine the optimal gap size for regeneration (Gutierres and Cabral pers. comm.). However, for reasons of expense, there is no maintenance (cleaning, pruning, thinning) of the young mahogany trees and their survival is uncertain. In addition, growth rates are thought to be too slow for the proposed 25 year cutting cycle. Snook (1993) suggests that the spacing of seed trees is more important than the minimum cutting diameter. Her recommendations include the maintenance of an even distribution of at least one seed tree per 3 hectares and delay of harvesting until after seed dispersal.

With a cutting cycle long enough to permit recovery of mahogany stocks, the application of simple cutting restrictions could conceivably constitute a crude silvicultural system. However, Gullison (in press) argues that use of minimum cutting diameters is inappropriate. The rarity of regeneration (disturbance) events in natural forest means that all mahogany trees in natural stands may be above the specified diameter. In any event, the low rates of regeneration observed in logged forest and the weak enforcement of existing restrictions means that current measures are incapable of maintaining long-term production of natural mahogany.

13.1.2 Research scale management

An early attempt to design a proper system of management for mahogany in natural forest was made by colonial foresters in the forests of Belize. The work involved a series of 'improvements' targeted specifically at the mahogany (see Plate 13.1 and Box 13.1). Silvicultural improvement was carried out over relatively large areas, over 1000 ha in Silk Grass Forest Reserve alone according to Lindo (1968), but the results showed that it was not a commercially viable proposition.

Plate 13.1. Mahogany seedling improvement in Belize (British Honduras). Note the extensive cleaning of the understorey leaving a high shelterwood. High Ridge, Belize. *Photo by N. Stevenson (1927), courtesy of the Commonwealth Forestry Association.*

Box 13.1. Silvicultural improvements in natural mahogany forests of Belize

Silvicultural work began in the natural mahogany forests of Belize in the 1920s, after it was suggested that the growth rates of young mahogany trees could be easily improved by freeing existing trees from climbers and other competing trees (Hummel 1925). Work focused on logged natural stands containing a relatively low density of mature mahogany trees in Silk Grass, Sibun Stann Creek, Columbia River and Freshwater Creek Forest Reserves (Oliphant 1925, Stevenson 1927b). In Silk Grass, average stocking of mahogany trees over 15 cm in dbh was 10 per hectare at the time (Stevenson 1927b).

The method that was first employed involved clearing a 'cylinder' around each mahogany seedling, approximately 2 m in diameter and 2 m in height, while leaving the natural forest canopy intact. This treatment was supposed to liberate the seedling, while protecting it from insolation, drought and insect attack by maintaining the forest canopy (Oliphant 1925). However, the mahogany seedlings failed to respond, presumably because of insufficient light.

After extensive experimentation, a more thorough, though more labour-intensive system was developed (Oliphant 1925). The system involved a number of silvicultural treatments, which are described by Stevenson (1927c):

Box 13.1. (continued)

- **Tree improvement.** Climbers were removed from large mahogany and other valuable timber trees. Inferior species which hindered crown development of valuable species were removed by girdling. These treatments encouraged seed production from valuable canopy trees and the establishment of natural regeneration.
- **First seedling improvement.** A considerable amount of work was required:
  a) opening of the canopy by cutting or girdling all unreserved species with a dbh of over 20 cm;
  b) complete removal of the middle strata of forest growth; and
  c) cleaning out of the undergrowth, leaving mahogany and other valuable seedlings.
  The considerable extent to which the forest was cleared is shown in Plate 13.1. Girdling was preferred to felling (Oliphant 1925): the gradual increase in light intensity under a girdled tree was thought to expose the young seedlings to less water stress than the sudden increase in light intensity caused by felling. All mahogany seedlings were marked with a stake and freed from undergrowth. The stakes were placed about 1 m from the mahogany seedlings so that climbers on the stakes did not interfere with the seedlings. Greater illumination allowed rapid growth of mahogany, while the shelterwood of valuable species provided the required protection from desiccation.
- **Reimprovement and tertiary improvement.** The cleaned understorey was treated again and sometimes a third time until a sufficiently stocking density had been obtained. These treatments freed established seedlings from creepers and enabled new seedlings to be located.

Once a suitable stocking had been obtained, associated regrowth was encouraged rather than cleaned, except where it exerted a suppressing influence. Seedlings that had been attacked by the shoot borer soon recovered their form after the dense regrowth had grown up (Stevenson 1929) and the regrowth was thought to limited further damage. The shoot borer was found to attack flushing leading shoots soon after the onset of rain and was most serious where the seedling was not protected by a shelterwood. 'With too drastic removal of overhead cover, damage by insects is severe, whereas where the cover is dense the seedlings remain in a condition of stagnation of which the small size and deep green colour of the leaves is symptomatic. An irregularly broken canopy seems to give the best results' (Oliphant 1926).

The ultimate aim was to produce desirable mahogany trees in all dbh classes, which would be harvested on a 10 year cycle (Stevenson 1944b). The predicted yield was 2.5 merchantable mahogany trees per hectare every 10 years and a selective cut of various other hardwoods. The size of a merchantable mahogany tree was not specified, although at the time a minimum felling diameter of 72.5 cm was in force (Stevenson 1944b). The proposed system of improvement therefore resembled a selection system, although the need for light at ground level to stimulate growth of mahogany seedlings would have led to a rather more broken canopy than is usually associated with the selection system.

Using this system Stevenson (1927c) records impressive early results. An increase of 390% in the number of mahogany seedlings in a 'typical' compartment in Silk Grass was obtained over a 17 month period between the first seedling improvement and reimprovement. Between reimprovement and tertiary improvement, an 80% increase was recorded. By 1929 the average stocking of 2 of the 4 treated compartments was 280 seedlings per hectare (Stevenson 1930).

However, by 1931, all silvicultural work had been abandoned owing to the Depression (Stevenson 1935) and the quality of the treated forest quickly deteriorated. After only 5 years from the 1931 improvement the regeneration was found in very disappointing condition with many previously healthy stems strangled by creepers and vines (Stevenson 1936). Seedlings in the 0-3 m class suffered highest mortality, although once a mahogany sapling passed into the pole stage, its continued development was fairly secure (Kinloch 1938). Because they stood for many years, girdled trees became festooned with creepers and caused serious damage to natural regeneration when they finally fell. The problem led Stevenson (1936) to condemn the policy of girdling of large trees, despite the problems associated with felling described by Oliphant (1925). However, had the forest been continuously managed, as was originally intended, creepers would have been cut from girdled trees and the damage they caused would probably have been less severe.

Improvement was discontinued in most forest reserves, although intermittent small-scale cleanings were carried out in Silk Grass in 1934 and again in the late 1930s and early 1940s (Stevenson 1935, 1940, 1942). However, it was recognised that such intensive treatment over a large area was not economically feasible (Stevenson 1942). Improvements were finally abandoned after a devastating hurricane, which struck Silk Grass in 1942.

There has recently been a resurgence of research in the now depleted natural mahogany forests of Belize and a number of ongoing experiments merit description.

Bird (pers. comm.) proposes the use of selective logging techniques in Belize and established a number of plots in 1994 to assess the feasibility of this approach. The aim of the experiment is to investigate the possibility of obtaining an economically viable timber harvest which also provides conditions suitable for natural regeneration of merchantable species, including mahogany. Treatments involve the removal of 6 merchantable stems per hectare over 40 cm dbh, compared to the traditional 0.25 stems ha^{-1} (Brokaw *et al.* 1995). However, there is a maximum girth limit (around 100 cm) to preserve the structure of the forest (Bird pers. comm.) and trees are selected on the basis of silvicultural criteria rather than minimum diameter (Bird 1994). The proposed length of felling cycle is 40 years. It is hoped that the proposed system will maintain biodiversity, although the forest composition will be changed by logging. No liberation, tree planting or climber cutting has been proposed and it remains to be seen if the gap sizes are big enough to encourage the growth of mahogany seedlings. The selective logging technique has not been designed specifically for mahogany production since most of the mahogany stems were removed previously (current stocking in the vicinity of the plots is only 4-5 trees over 10 cm dbh per hectare).

Liberation thinning plots have been set up in the Rio Bravo Conservation and Management Area (RBCMA) and Chiquibul forests of Belize to demonstrate how productivity, quality and stocking of commercial timber species may be improved (Brokaw *et al.* 1995, Martens pers. comm.). Diagnostic sampling is used to select leading desirables and thinning carried out to liberate them from competition (Hutchinson 1991). Mahogany and other species of commercial value of all age and size classes will be improved in this way (Martens pers. comm.). It is hoped that the favourable response of mahogany and other species to liberation will permit a reduction in cutting cycle.

Both liberation thinning and selective logging techniques have only recently been applied to natural mahogany forest and it will be many years before the results of the experiments are known. The exponents of the different techniques believe that they may be applicable at the commercial scale, but the lack of associated disturbance limits the potential for establishment of mahogany regeneration and thus the development of mahogany-rich natural forest (Brokaw pers. comm.).

Programme for Belize (a non-governmental organisation) has recently initiated a sustainable timber extraction programme in the RBCMA, designed to encourage natural regeneration of mahogany (PfB 1996). A 40 year felling cycle has been proposed. Before logging operations commence, a detailed stock survey is carried out and seed trees are identified. Under current harvesting guidelines, at least 20 mahogany seed trees per 100 ha (including the single largest mahogany tree) and all mahogany trees under 50 cm in dbh must be retained (PfB 1997). Directional felling techniques are used (see Plate 13.2). The planned haresting intensity is low (an average of 1.6 stems ha^{-1} for all species in the 1997 cut) and disturbance of both canopy and undergrowth is considered to be insufficient for mahogany regeneration. Therefore, 'sector cuts' (wedge-shaped, clear-felled patches) are made down-wind of mahogany seed trees to allow more light to reach the forest floor and regenerating seedlings. Subsequent silvicultural operations (weeding, cleaning and thinning) are thought to be unnecessary, although given earlier experience in Belize (see Box 13.1) this may be a risky assumption. The programme is still in its experimental stages, but has recently been certified and is expected to evolve into a commercial scale operation.

Plate 13.2. Directional felling of mahogany in natural forest. Rio Bravo Conservation and Management Area, Belize.

13.2 Silvicultural systems in plantations

Experience in the use silvicultural systems in mahogany plantations has recently been reviewed by Mayhew *et al.* (in press), from which the following account is taken.

Some silvicultural systems designed for species-poor temperate forests of western Europe may offer appropriate models for management of mahogany plantations. These include the clear-cutting system, the shelterwood system and the selection system, each of which has a number of variations (see Troup 1928, Smith 1986 or Matthews 1992 for more details).

The silvicultural system commonly applied to plantations of most tropical species involves clear-cutting and replanting or coppicing to establish the next rotation. Clear-cutting and replanting has been carried out in mahogany plantations in Martinique with good results (Chabod in press) and it is likely that many mahogany plantations elsewhere will be managed in a similar manner. However, the clear-cutting system is not designed to encourage establishment and growth of natural regeneration under mature trees (although the system can be adapted to make use of advance growth).

Silvicultural systems that are designed to encourage natural regeneration have a number of economic advantages, including reduced nursery, transportation and planting out costs. This section therefore discusses the potential of natural regeneration in mahogany plantations and includes case studies of silvicultural systems which make use of natural regeneration.

13.2.1 Natural regeneration in mahogany plantations

Mahogany sets seed at an early age (see Section 4.1). Well-spaced stands reproduce earlier than closely spaced stands, where trees crowns are competing for light. Good seed years are frequent but irregular, occurring on average about twice in every 3 years in Fiji (Streets 1962). In Sri Lanka a good seed year usually occurs once in every 2-3 years.

In most cases, mahogany trees will have been seeding for several decades before they reach maturity and are ready for felling. Profuse natural regeneration is characteristic of mahogany stands in Trinidad (Marshall 1939), the French Antilles (Anon. 1959), Indonesia (Alrasjid and Mangsud 1973), the Solomon Islands (Chaplin 1993), the Philippines, Fiji, St. Lucia and Sri Lanka (Mayhew *et al.* in press) (see Plate 13.3). In many instances, mahogany regeneration has established under a closed canopy where light availability is low. Seedlings can apparently survive for many years in these conditions but will not grow until light availability increases (see Section 2.5.3). Many plantations therefore contain a bank of natural regeneration, waiting for the removal of the canopy. With an appropriate silvicultural system, this natural regeneration can be used to establish the next rotation.

Plate 13.3. Dense natural regeneration of mahogany in a selectively logged stand. Welikande plantation, Sri Lanka.

The existence of mature mahogany trees does not guarantee the presence of natural regeneration of mahogany in the understorey. In St. Lucia, Sri Lanka and the Philippines some mahogany stands are notably free of regeneration, even though nearby stands may have high densities. It has been pointed out that seedlings in the more accessible stands are often uprooted by the general public for garden planting (Pablo pers. comm., Tilakeratne pers. comm.). However, not every case of regeneration failure can be explained by seedling removal. Unfortunately, there have been no documented surveys of natural regeneration in mahogany plantations and suggested explanations for its presence/absence given below are unproven.

Muttiah (1965) noted that in certain Sri Lankan plantations, roadsides and ridges were the only sites that had been colonised, although older, pole-stage regeneration was present throughout. He suggested that the deep litter layer was preventing the establishment of regeneration thoughout undisturbed parts of the plantations and that the pole-stage trees were established after disturbance during an earlier selective felling. Parraguirre (pers. comm.) believes that deep litter is preventing natural regeneration under experimental plantations in Mexico. Site factors may also be important. Bobb (pers. comm.) suggests that the lack of regeneration on one of two similar sites in St. Lucia may be due to poor drainage and waterlogging after rains. Streets (1962) observed that mahogany regenerates most profusely in Trinidad on calcareous clays and fertile soils, where moisture is adequate all year round. Wolffsohn (1961) demonstrated that insect predation of mahogany seeds was preventing regeneration of mahogany in natural forests of Belize; an explanation which may also be relevant to plantations. Low light intensity (Muttiah 1965) and root competition from mature mahogany trees (Coster 1935) may also be factors, but it must be remembered that regeneration has been observed under conditions of both low light intensity and high stand density (e.g. in Fiji and Sri Lanka) and other factors may be more significant. Low rates of seed production and poor seed viability are other possibilities.

Mayhew *et al.* (in press) note that natural regeneration under *closed* plantation canopies in many countries is remarkably free of shoot borer damage, though attack has been found to occur at plantation *edges*. Likewise, Whitman (no date) notes an absence of attack of natural regeneration in St. Lucia, suggesting that this may be due to the shade of the overstorey (although the mechanism preventing damage has yet to be identified). The lack of attack is surprising, given the quantity of easily accessible shoot material on which the shoot borer could lay its eggs. It is another reason why forest managers may wish to consider the use of a silvicultural system other than clear-cutting and replanting.

13.2.2 Systems utilising natural regeneration of mahogany

Until recently there has been only one attempt to adopt a silvicultural system which involves the use of natural regeneration in mahogany plantations. The experience of the Sri Lankan Forest Department with the single tree selection system in the 1950s is described below (see Section 13.2.2.1). Over the last few years, a modified uniform shelterwood system has been designed for Sri Lanka (see Section 13.2.2.2) and a group selection system has been proposed for mahogany plantations in St. Lucia (see Section 13.2.2.3). However, at the time of writing, plans for adoption of recently designed systems had not been fully implemented.

13.2.2.1 A single tree selection system in Sri Lanka

Mahogany was first planted in Sri Lanka in 1889 (Tissverasinghe and Satchithananthan 1957). The species was mixed with *Artocarpus integrifolia*, *Tectona grandis* and local timber species, including *Chloroxylon swietenia*, *Mesua ferrea* and *Pericopsis mooniana*, in what was later to become known as Sundapola plantation. The successful growth of the mahogany led, between 1925 and 1960, to the gradual establishment of many other plantations which were principally mixtures of mahogany and *A. integrifolia* (Sandom and Thayaparan 1995).

A single tree selection system was adopted for the management of mahogany in Sundapola plantation in 1954 in order to maintain its 'scientific value' (Tissverasinghe and Satchithananthan 1957). The selection system was considered most suitable because of the presence of natural regeneration under the maturing stands, which had already begun to develop into pole-stage trees. Diameter class frequency histograms showed a shift from a typical even-aged (or normal) distribution to what resembled a selection (or de Liocourt) distribution (Tissverasinghe and Satchithananthan 1957, Muttiah 1965). Muttiah (1965) believed that the selection system was more productive than uniform systems, citing an annual percentage increment in volume ranging from 5.6% to 9.2% (note that these figures were taken from measurements in stands *in the process* of conversion from even-aged clear-cut system to an uneven-aged selection system). Other advantages offered by the selection system include flexibility for forest managers, steady timber yields, negligible establishment costs, low suseptibility to pests and diseases and aesthetic values. However, doubts over the suitability of a selection system were also expressed and problems such as 'expense, lack of properly trained staff and the ease with which selection can degenerate into uncontrolled exploitation' were noted (Tissverasinghe and Satchithananthan 1957).

Sundapola plantation was carefully managed under the single tree selection system for a decade. The system may have been applied to other mahogany plantations, though due to loss of most plantation records this cannot be verified. However, at some stage after the mid-1960s, the fears of Tissverasinghe and Satchithananthan were realised. The selection system degenerated into systematic (albeit low-intensity) high-grading which subsequently depleted many of the mahogany stands of plus trees and a few stands of most of their mahogany stock. It appears that much of this work was carried out in the mistaken belief that it constituted correct selection system silviculture. The selection system failed because it had assumed a detailed understanding of complex forestry principles and an intimate knowledge of forest structure by all (even the most junior) forest officers (Sandom and Thayaparan 1995). The high levels of training and continuity of service required were (and still are) unavailable to most Forest Department staff.

Plate 13.4. An uneven-aged mahogany stand created by selective logging. Marukwatura plantation, Sri Lanka.

The fundamental difficulty with the adoption of the single tree selection system is its unsuitability for light-demanding species such as mahogany. In Sri Lanka, selection system management was not carried out for a sufficient length of time for this drawback to become evident. However, in uneven-aged, single tree selection stands, natural regeneration is expected to grow under relatively small openings in the canopy for long periods of time (see Plate 13.4). Experience in natural forests indicates that mahogany does not grow well under such conditions and the system is much better suited to shade-tolerant species. Although, in theory, a stand can be thinned to promote growth of a light-demanding species, in practice the amount of work involved make such treatments unrealistic. This realisation led to the adoption of an interim management plan in 1994 advocating a different silvicultural system (Sandom and Thayaparan 1994).

13.2.2.2 A modified uniform shelterwood system in Sri Lanka

Because of the shortcomings of the selection system, plans have been developed to manage mahogany plantations in Sri Lanka under a uniform shelterwood system (Sandom and Thayaparan 1995). The shelterwood system has not been applied to mahogany plantations before, although the system has been considered in the French Antilles (Anon. 1959, Tillier 1995) and Fiji (Singh pers. comm.). The aim of the system is to convert complex (mixed and multi-storied) mahogany stands into simpler, more productive and more easily managed stands (see Figs. 13.1a-h). Natural regeneration is encouraged by a regeneration felling and remaining seed trees are removed 5 years later. The system is suitable for a light-demanding species such as mahogany because trees are exposed to conditions of full light throughout most of the rotation.

The first stage of the shelterwood system is the regeneration felling (see Fig. 13.1b). This will be carried out in stands which are mature or which have a number of mature trees. As well as encouraging the growth of existing advance regeneration, the regeneration felling will aid the establishment of new seedlings. Ideally, regeneration felling should coincide with good seed years, after the seed has ripened but before the onset of rains when the seed starts to germinate. Opening of the canopy is likely to stimulate strong weed growth (Matthews 1992) and if natural regeneration is not established quickly, foresters will be faced with the daunting task of cutting back secondary regrowth until the next good seed year.

In the absence of regeneration trials, it is difficult to decide how many seed trees to retain. The shelterwood must be sufficiently dense to ensure that all parts of the forest floor are exposed to falling seeds. In the Sri Lankan Management plan a density of 35 stems ha^{-1} is proposed. In contrast, forest managers in Fiji believe that 80-100 trees will be required to control weedy regrowth and to produce sufficient natural regeneration (Singh pers. comm.), although it is worth noting that mature tree crowns in Fiji are smaller than those in Sri Lanka, owing to the shorter (35 year) rotation length. The main disadvantage of a dense shelterwood is the damage to established natural regeneration caused by the seed tree felling (see Fig. 13.1d). Alrasjid and Mangsud (1973) measured the quantity of natural regeneration under a 28-year-old mahogany stand in Indonesia before and after harvesting. During felling operations, 65% of seedlings and 34% of saplings were destroyed. Although this survey was carried out under a closed rather than shelterwood canopy, it illustrates the damaging effect of forest operations on natural regeneration. However, a low density shelterwood combined with low impact harvesting and extraction techniques will minimise damage (Sandom and Thayaparan 1994).

Fig. 13.1. Modified uniform shelterwood system in Sri Lanka

a) Present situation: stands are uneven-aged; mahogany stocking is variable and other species are present in varying proportions

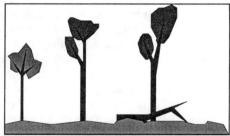

b) Year 0: **regeneration felling**; selected seed trees are left including other key species, e.g. *Artocarpus integrifolia*; 35 stems ha^{-1}

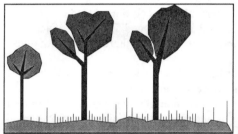

c) Years 0-5: **maintenance**; undergrowth is weeded to encourage natural regeneration

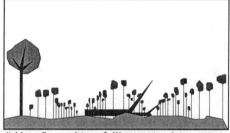

d) Year 5: **seed tree felling**; natural regeneration reaches a high density; >2500 stems ha^{-1}

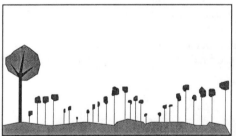

e) Year 8: **singling and spacing** of natural regeneration (transplantation if necessary); some other species are retained; 2500 stems ha^{-1}

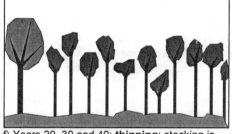

f) Years 20, 30 and 40: **thinning**; stocking is gradually reduced; 900, 600 and 270 stems ha^{-1} respectively

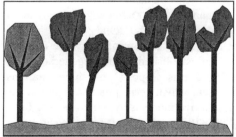

g) Year 50: **final thinning**; final crop trees are well spaced; 100 stems ha^{-1}

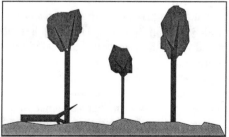

h) Year 60: **regeneration felling**; the cycle is repeated from c)

A number of thinnings will be carried out during the course of the rotation. Initially stands will be subject to a systematic thin to space the natural regeneration (see Fig. 13.1e); subsequently stands will be selectively thinned to take out poorly formed stems and expand the growing space of residual trees (see Figs. 13.1f and g). The proposed thinning regime is based on predicted dbh growth rates of just under 1 cm yr^{-1}, but no growth and yield data is available from which to determine *optimal* thinning frequency and intensity (see Section 8.5.1 for proposed Sri Lankan regime). There is a good local market for thinnings down to 15 cm, for the local furniture industry, and demand for logs and poles less than 15 cm in diameter for construction, fencing and farm implements.

An important aspect of the shelterwood system is early selection of seed trees. Crown development takes time and cannot be achieved by a single, last-minute thinning (Matthews 1992). It is therefore important to thin sufficiently during the second half of the rotation. In the design of an untried shelterwood system for mahogany plantations in the French Antilles, Anon. (1959) recommended 3-4 selective cuts towards the end of the rotation, removing a third to a quarter of the timber volume each time. Matthews (1992) stresses that at least half of the crop yield should be produced from thinnings. The thinning regime proposed (see Figs. 13.1f and g) is in broad agreement with these recommendations.

The shelterwood system has a number of attributes which make it particularly suitable for Sri Lankan plantations. The system permits the development of clear, straightforward management guidelines for forest officers, avoiding the complexities and subjectivity of the selection system. Continuous forest cover is maintained, reducing the risk of soil erosion and allaying widespread local fears of permanent forest clearance. It is also possible that the shelterwood will reduce the risk of shoot borer attack to natural regeneration (see Section 11.1.2.3). However, there are disadvantages. The system has been modified to permit the retention of *Artocarpus integrifolia* (jak), an important source of fruit for local communities, and the conservation of local timber species (e.g. *Chloroxylon swietenia* and *Vitex pinnata)*. The need to accommodate non-mahogany species, will adversely affect plantation productivity and complicate prescribed silvicultural operations in certain stands.

13.2.2.3 A group selection system in St. Lucia

St. Lucia (in the Lesser Antilles) has a small mahogany resource of just over 100 ha (Andrew 1994). Mahogany was introduced as an ornamental species in the 1930s, although plantation establishment did not begin until the 1960s. Mahogany was established in pure plantations, in a taungya system with bananas (which later become pure) and in a mixture with *Hibiscus elatus* (blue mahoe), with mahogany typically at an initial 3 x 3 m spacing. Thinning has been erratic, with some stands heavily crown thinned and others virtually untreated (Whitman no date). Recurrent hurricane damage has created gaps in the canopy, stimulating growth of established natural regeneration.

Although no management plan has been designed, it is proposed that the plantations are managed under a variation of the selection system which corresponds roughly to the group selection system. The sequence of silvicultural treatments is hypothetical (see Figs. 13.2a-g). It is anticipated that thinning and felling will be carried out on a 7 year cycle, with a rotation length of about 35 years based on existing growth data. Where blue mahoe is present in the stand, it will be maintained at a level of 15% of total stocking to encourage the development of tall, straight mahogany stems.

Fig. 13.2. Group selection system in St. Lucia

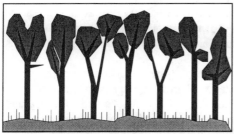

a) Present situation: mahogany stands becoming uneven-aged and multi-storied

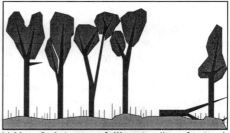

b) Year 0: **1st group felling**; tending of natural regeneration in first group

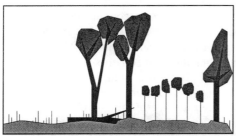

c) Year 7: **2nd group felling**; tending of natural regeneration in 2nd group; spacing of natural regeneration in 1st group

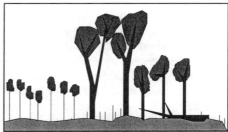

d) Year 14: **3rd group felling**; tending of natural regeneration in 3rd group; spacing of natural regeneration in 2nd group; thinning in 1st group

e) Year 21: **4th group felling**; tending, spacing and thinning in groups repeated

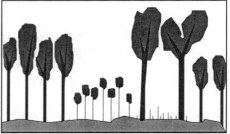

f) Year 28: **5th group felling**, tending, spacing and thinning in groups repeated; natural regeneration from 1st group now nearing maturity

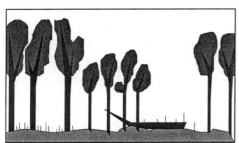

g) Year 35: **6th group felling** (natural regeneration from 1st group); stand now contains 5 different age groups; felling continued on a 7 year cycle

Many details, including tending and thinning regimes and the desirable size of groups, have yet to be ascertained. To be suitable for a light-demanding species such as mahogany, felling groups may need to be large (perhaps over 500 m² or at least 5 mature trees), although much will depend on the height of the canopy in stands adjacent to the group. Productivity will be low if felling groups are small and natural regeneration is shaded by surrounding trees.

One of the difficulties with any selection system is harvesting and extracting trees without damaging the remaining trees in the stand. If group felling is carried out, it is possible to fell selected trees towards the middle of the group to avoid catching and snagging surrounding trees. In St. Lucia, mechanised extraction is forbidden and contractors use portable sawmills to saw up cut logs *in situ*. Planks are carried out by foot and the amount of damage caused to remaining trees is therefore minimised.

The group selection system is considered to be the most suitable system for mahogany plantations in St. Lucia because it is able to accommodate key local constraints on management practices. The need to protect watersheds is considered of paramount importance by the Forestry Department (James pers. comm.). Much of the mahogany has been planted on steep slopes where the threat of soil erosion during forest operations is severe and the adoption of the group selection system will minimise the risks. The system is also appropriate for the management of hurricane-damaged stands. Regeneration in pockets of damage can be used to create new groups. Forestry Department staff are sufficiently well trained to put such a complex system into practice, but the demands on staff time will significant if the system is to be made to work, despite the small size of the mahogany resource.

13.2.2.4 Application of silvicultural systems in other countries

From the work that has been carried out in Sri Lanka and St. Lucia, it appears that the uniform shelterwood and group selection systems are suitable for the long-term management of mahogany plantations, although it will be many years before a quantitative assessment is possible. In theory, any silvicultural system may be applied to mahogany plantations, as long as trees receive sufficient light. However, the intensity of management required to achieve suitable conditions of light under the single tree selection system, makes it an unrealistic choice.

The successful adoption of a silvicultural system does not depend simply on finding a system which meets the light requirements of mahogany. Forest managers need to take careful note of important local constraints on plantation management practices (see Table 13.1). Failure to take local constraints into account is likely to result in the failure of the adopted system, as was demonstrated by the single tree selection system in Sri Lanka. In general, the shelterwood system appears to be the most appropriate system for the long-term management of mahogany plantations.

Table 13.1. The suitability of silvicultural systems under different constraints on plantation management: suggestions based on work in Sri Lanka and St. Lucia (Mayhew et al. in press)

Type of constraint	Examples of constraints	Clear-cut system	Shelterwood system	Group selection system
Environmental	Soil erosion is a high risk; conservation of native species found in plantations is a high priority; hurricane damage is a high risk	Not suitable	Suitable with modification	Suitable
Social	Production of benefits for local communities is a high priority	Not suitable	Suitable with modification	Suitable
Institutional	Silvicultural malpractice (due to an inadequate, poorly managed and/or untrained workforce) is a high risk	Suitable	Suitable	Not suitable
Economic	Maximising stand profitability is a high priority	Suitable	Suitable	Not suitable

13.3 Summary

- Owners and concessionaires have tended to exploit rather than manage natural mahogany forests. The imposition of cutting restrictions and retention of seed trees has been prescribed in some cases, but such techniques have not secured sufficient natural regeneration and the chance of future mahogany harvests in such areas seems remote.
- It is difficult to be specific about the kind disturbance which provides conditions most favourable for mahogany regeneration. However, there is now much evidence to indicate that selective logging of mahogany in natural forest, which produces small gaps in the canopy and a relatively undisturbed understorey, greatly limits establishment and growth of regeneration.
- Experimental management of natural mahogany forest in Belize demonstrated that thinning of the canopy combined with careful weeding and cleaning of the understorey around mahogany seedlings greatly enhanced the density of natural mahogany regeneration. Stands required continuous maintenance: when tending was abandoned, many seedlings died and growth rates of survivors slowed.
- Some researchers now advocate the clear-cutting of patches around seed trees, to create similar conditions to those produced by the kind of natural disturbances that favour mahogany regeneration. The arguments for the use of such techniques are convincing, but it will be many years before useful results are available from trials.
- The apparent ease with which mahogany regenerates in many plantations gives managers the potential to change from the traditional silvicultural system of clear-cutting and replanting to a system which makes use of existing natural regeneration. There are some important benefits to be gained from use of natural regeneration, including reduced nursery costs, enhanced watershed protection and the possibility of reduced shoot borer attack.
- As in natural forest, natural regeneration in plantations is unable to grow beyond the sapling stage until the canopy has been opened. Silvicultural systems, such as the single tree selection system, which condemn young mahogany trees to a shaded

existence for much of the rotation may, in the long term, seriously reduce stand productivity.
- Two silvicultural system which appear to be appropriate for mahogany plantations are proposed. The *uniform shelterwood system* involves a regeneration felling to produce a low density canopy or shelterwood (perhaps 30-50 stems ha^{-1}) about 5 years before clear-felling. The *group selection system* involves the removal of canopy trees in groups (the optimal size of canopy opening has yet to be determined).
- The successful adoption of a silvicultural system depends, to a large extent, on factors which constrain management activities and decisions. The existence of institutional and economic constraints greatly increase the risk of failure of complex selection systems. Similarly, ecological, environmental and social constraints make simple clear-cutting systems unsuitable. In general, variations of the shelterwood system appear to offer the best solution.

Chapter Fourteen

Conclusions

Mahogany is one of the most valuable and economically important tropical timbers. However, almost all the timber currently traded is obtained from unmanaged natural forests. The uncontrolled exploitation of mahogany throughout its natural range has led to the degradation of many natural stands and raised serious concerns amongst conservationists. In many areas, particularly in Central America, mahogany is now considered to be commercially extinct. Even in the more extensive forests of Amazonia, stocks of mahogany may be logged out within a few decades if extraction continues at current rates (de Barros *et al.* 1992, Rodan *et al.* 1992, CITES 1997). The development of sustainable management approaches is essential if the timber resource is to survive (Newton 1993).

Recent research on the regeneration ecology of mahogany has emphasised the importance of major canopy-level disturbance, caused naturally by hurricanes, fire or flooding, in creating suitable conditions for seedling establishment. The lack of regeneration following the more limited disturbance associated with logging poses a significant silvicultural challenge for the management of natural mahogany forests. A new approach, currently being tested in Belize, involves the clear-cutting of patches near mahogany seed trees to create similar conditions to those produced by the kind of natural disturbances that favour mahogany regeneration. However, the disturbance associated with patch cutting or intensive logging around seed trees may compromise other objectives of sustainable forest management, particularly when the cost of such interventions is taken into account (Rice *et al.* 1997, Gullison in press).

Given the problems with natural forest management, the development of mahogany plantations to secure future supplies of the timber and perhaps to lessen the pressure on natural forests merits serious consideration. In many respects, mahogany is an ideal plantation species. Seedlings are easy to grow in the nursery. Nursery managers have considerable flexibility over the choice of planting stock and even direct seeding has produced satisfactory results. Mahogany is tolerant of a wide range of soil types and can be established successfully on rocky or steeply sloping sites in moist conditions. The suitability of mahogany as a plantation species has led to the establishment of some 200,000 ha of plantations world-wide, with extensive areas in Indonesia, Fiji and the Philippines. However, mahogany plantations in many tropical countries have not been sufficiently productive to attract further investment. There are several factors which explain the low productivity observed; there are also a number of ways in which productivity could be enhanced.

In most cases, plantations have been established without a prior programme of provenance testing. However, readily available mahogany seed may well be relatively ill-adapted to available planting sites. The identification of suitable provenances will produce substantial benefits in plantation productivity through improvement of tree form and growth rate, and provenance testing should therefore be given highest priority. Given the probable narrow genetic base of many established plantations, managers should aim to test a wide range of mahogany provenances from natural origin and compare them with locally available seed supplies. It is to be hoped that the network for the genetic resources of mahogany, recently established by the FAO (Patiño 1997), will facilitate the dissemination and testing of germplasm in areas where plantations are being established.

The most significant problem facing mahogany plantations has been the mahogany shoot borer (*Hypsipyla* spp.). Low branching and loss of the main leader caused by damage to the apical shoot of young trees can drastically reduce future timber production. There are a number of preventative and remedial silvicultural techniques which can be used to limit shoot borer damage. Given that mahogany provenances vary in their resistance to shoot borer attack, the selection of the most resistant provenance may be an effective preventative measure, further emphasising the importance of provenance trials. Techniques which assist recovery after attack, by promoting vigorous vertical growth and restricting the development of lateral branches, are recommended. Examples include careful site matching, pruning, and managing a stand to create conditions of lateral shade and overhead light. Evidence indicates that, with careful plantation planning and timely intervention, well-formed trees can be produced in situations where the shoot borer poses a serious threat. The real problem facing forest managers is the high cost of intensive management over the early part of the rotation. In regions prone to high incidence of shoot borer attack, adoption of the range of silvicultural techniques required to minimise damage may incur unacceptable costs.

The health and productivity of many mahogany plantations around the world has suffered severely as a result of poor maintenance. Mahogany is a light-demanding species and requires careful management to ensure that the crowns of individual trees in the stand receive direct overhead sunlight. On open sites, pure mahogany stands grow rapidly, but mixed plantations need careful management to ensure that non-mahogany species do not close canopy above the mahogany. Appropriate species are faster growing, but with a shorter rotation length than mahogany. If mahogany is planted in lines through existing natural forest, it is important to ensure that the residual forest canopy has been removed and that understorey vegetation does not close over the line. To maintain optimal conditions for growth, all stand types require selective thinning at regular intervals throughout the rotation. It is no coincidence that the most productive mahogany plantations (e.g. those in Martinique) are also those in which high standards of silviculture have been observed.

Although productivity is an important means of enhancing the profitability of mahogany plantations, there are other factors to be considered. Densely stocked mahogany stands producing large volumes of small diameter timber may be extremely productive but relatively unprofitable if there is no market for thinnings. Before a plantation programme is initiated, an assessment of local and international timber markets is essential. Only a few countries appear to have satisfactory local markets for small diameter mahogany timber, although international 'niche' markets may be open to producers of certified timber. Where markets for mahogany thinnings do not exist, it may be advisable to plant mahogany in a mixture with a marketable fast-growing timber species. The latter can be thinned out early on in the rotation and may help to reduce damage by shoot borers during the critical period of early growth for mahogany trees. Alternatively, mahogany may be planted at low density in lines through natural forest.

Once established, the long-term prospects for mahogany plantations are good. The profuse natural regeneration found under many mature mahogany plantations may be used to establish the next crop, reducing nursery and plantation establishment costs. Certain silvicultural systems, such as the uniform shelterwood and group selection system, appear to be suitable for the long-term management of such mahogany plantations. Adoption of a silvicultural system using natural regeneration may bring environmental and social benefits not available to practitioners of the traditional clear-cutting and replanting system, and should be considered by those committed to sustainable plantation management.

Reported studies of the economic viability of mahogany plantations are too few to merit a separate section in this book. However, it is recognised that, to a large extent, economic considerations determine the kind of silvicultural practices that are adopted. Browder *et al.* (1996) studied discounted cash-flows for mahogany planted in degraded fallow, agroforestry systems and pure stands in Brazil. Pure stands were more profitable than the alternatives, with 9 out of 10 of the scenarios investigated demonstrating financial viability. If ecological constraints (such as the shoot borer) are ignored, Browder *et al.* concluded that the financial prospects for mahogany silviculture were good. One of the main factors hindering efforts to grow mahogany in plantations is the underpricing of large trees extracted from natural forest (Rodan *et al.* 1992). Natural mahogany trees have an average volume per harvested tree of 5.4 m^3 and timber that is regarded as superior to that of plantation mahogany by most timber processors. However, as natural supplies become scarcer and mahogany plantations become more productive, plantation profitability will increase. If current demand for the timber continues, establishment of mahogany plantations will become a more attractive economic proposition in many countries.

Over 30 years ago, in his treatise on the ecology and management of mahogany, Lamb (1966) came to the following conclusion: 'I have seen sufficient examples of mahogany grown in both natural reproduced stands and in plantations under conditions approximating forest environment [i.e. line planting in natural forest] to convince me that mahogany species hold great promise as plant material of economic value. Land managers must be trained to recognize and maintain suitable environmental conditions for the satisfactory, if not optimum, development of mahogany'. Since Lamb's time, progress in mahogany silviculture has been slow and there is still much that can be done to improve plantation quality. It is hoped that this book will help forest managers to create the conditions required to enable mahogany to achieve its full potential in both plantations and natural forests.

Personal Communications

Anise, R. Forest Officer, Management Division, Department of Forestry. Colo-i-Suva, Fiji.
Appanah, S. Director. Forest Research Institute of Malaysia (FRIM). Kuala Lumpur, Malaysia.
Bird, N. Team leader, Forest Planning and Management Project, Overseas Development Administration. Belmopan, Belize.
Bobb, M. Forest Officer, Forestry Department. Castries, St. Lucia.
Bodley, M. Nursery manager, Forestry Department. Castries, St. Lucia.
Brokaw, N. Director of Tropical Forest Studies and Conservation, Manomet Observatory for Conservation Sciences. Massachusetts, USA.
Cabral, R. Técnico, Las Caobas Ejido. Quintana Roo, Mexico.
Castillo, E. Ecosystems Research and Development Bureau. College, Laguna, the Philippines.
Chabod, D. Directeur Régional, Office National des Forêts. Fort-de-France, Martinique.
Chand, M. Wood Technology Division, Department of Forestry. Colo-i-Suva, Fiji.
Cruz, M. Ingeniero Forestal, Lancetilla Gardens. Tela, Honduras.
Ennion, R. Research Forest Officer, Forest Department. Melinda, Belize.
Evo, T. Forest Officer, Silvicultural Research Division, Department of Forestry. Colo-i-Suva, Fiji.
Fox, J. Assistant Professor of Environmental Biology, Curtin University. Perth, Australia.
Francis, J. Research Forester, International Institute of Tropical Forestry. San Juan, Puerto Rico.
Gillies, A. Research Scientist, Institute of Terrestrial Ecology. Edinburgh, United Kingdom.
Gutierres, S. Técnico, Las Caobas Ejido. Quintana Roo, Mexico.
ul Hassan, Z. Team Leader of the Quintana Roo Forestry Management Project, Overseas Development Administration. Chetumal, Mexico.
Iakapo, M. Assistant Director of Forestry, Ministry of Agriculture, Forests, Fisheries and Meteorology. Apia, Western Samoa.
Iputu, S. Senior Forest Officer (Research), Ministry of Forests, Environment and Conservation. Forest Research Station, Munda, Solomon Islands.
Isles, M. Forest Manager, Kolombangara Forest Products Ltd. Kolombangara, Solomon Islands.
James, B. Chief Forester, Forestry Department. Castries, St. Lucia.
John, L. Environmental Officer, Forestry Department. Castries, St. Lucia.

Kamath, M. Senior Forest Entomologist, Silvicultural Research Division, Department of Forestry. Colo-i-Suva, Fiji.
Kotulai, J. Forest Officer, Rakyat Berjaya Sdn. Bhd., Yayasan Sabah. Luasong Forestry Centre, Sabah.
Kubuabola, T. Forest Officer, Silvicultural Research Division, Department of Forestry. Colo-i-Suva, Fiji.
Lapis, E. Ecosystems Research and Development Bureau, Department of Environment and Natural Resources. College, Laguna, the Philippines.
Lilo, J. Assistant Forestry Supervisor, Kolombangara Forest Products Ltd. Kolombangara, Solomon Islands.
Malvar, W. Forest Management Bureau. Nueva Vizcaya, Luzon, the Philippines.
Martins, P. Líder de Proyecto de Producción en Bosques Naturales, CATIE. Turrialba, Costa Rica.
Matsumoto, K. Forestry Division, Japan International Research Centre for Agricultural Science Project. c/o Yayasan Sabah, Kotakinabalu, Sabah, Malaysia.
Meza, R. Asistente Regional, Administratión Forestal del Estado Corporación Hondureña de Desarrollo Forestal. La Ceiba, Honduras.
Moce, N. Forest Officer, Management Division, Department of Forestry. Colo-i-Suva, Fiji.
Pablo, N. Ecosystems Research and Development Bureau, Department of Environment and Natural Resources. College, Laguna, the Philippines.
Palmer, J. Forestry Consultant, Tropical Forestry Services Ltd. Oxford, United Kingdom.
Parraguirre, C. Instituto Nacional de Investigaciones Forestales y Agropecuarias. San Felipe Bacalar, Quintana Roo, Mexico.
Patricio, H. Environmental and Resources Research Division. Tuguegarao, Luzon, the Philippines.
Ratnayake, District Manager, State Timber Corporation. Kurunegala, Sri Lanka.
Sandom, J. Forest Management and Plantation Project Team Leader, ODA. Colombo, Sri Lanka.
Sarshar, D. Technical Adviser, Concern. Dhaka, Bangladesh.
Seroma, L. Divisional Forestry Officer (South), Department of Forestry. Fiji.
Singh, K. Acting Principal Silviculturalist, Silvicultural Research Division, Department of Forestry. Colo-i-Suva, Fiji.
Somarriba, E. Líder de Proyecto, CATIE. Turrialba, Costa Rica.
Speed, P. Divisional Manager Poitete, Kolombangara Forest Products Ltd. Kolombangara, Solomon Islands.
Stanley, S. Research Project Manager, CATIE. Turrialba, Costa Rica.
Thayaparan, S. Assistant Conservator of Forests, Forest Department. Colombo, Sri Lanka.
Tilakeratne, D. Research Officer, Forest Department. Forest Research Station, Kumbalpola, Sri Lanka.
Timyan, J. Researcher, Haiti Tree Seed Centre. Haiti.
Tipper, R. Research Fellow, Institute of Ecology and Resource Management, Edinburgh University, United Kingdom.
Turnbull, C. Team Leader, Guyana Forestry Commission Support Project. Georgetown, Guyana.

Valera, C. Nursery Manager, Magat Reforestation Project. Nueva Vizcaya, Luzon, the Philippines.
Vidal, P. Forest Officer, Forestry Department. Castries, St. Lucia.
Villafranco, F. Forest Officer, Forest Department. Toledo District, Belize.
Viuru, X. Divisional Manager Ringgi, Kolombangara Forest Products Ltd. Kolombangara, Solomon Islands.
Wadsworth, F. Research Forester, International Institute of Tropical Forestry. San Juan, Puerto Rico.
Watt, A. Research Scientist, Institute of Terrestrial Ecology. Bush Estate, Penicuik, United Kingdom.
Wheatley, J. Director, Proyecto de Conservación y Mejoramiento de Recursos Forestales, ODA-CONSEFORH. Siguatapeque, Honduras.
Wilson, R. Programme Manager, Programme for Belize. Belize City, Belize.

References

Abe, K.I. **1983** Plantation forest pests in Sabah. Publication no. 8, Forest Research Centre, Sabah, Malaysia. 119 pp.

Abrams, E.M., Freter, A., Rue, D.J. and Wingard, J.D. **1996** The role of deforestation in the collapse of the Late Classic Copán Maya State. *In*: L.E. Sponsel, T.N. Headland and R.C. Bailey (eds.) *Tropical deforestation, the human dimension*, pp. 54-75. Columbia University Press.

Adesida, G.A. and Adesogan, E.K. **1971** The limonoid chemistry of the genus *Khaya* (Meliaceae). *Phytochemistry* **10**(8): 1845-1853.

Adesogan, E.K. and Taylor, D.A.H. **1970** Limonoid extractives from *Khaya ivorensis*. *Journal of the Chemical Society* **12**: 1710-1714.

Agbayani, W.C. and Marcelo, H.B. **1955** Preliminary study on the spot seeding of large-leaf mahogany in a logged-over area in Basilan. Occasional Paper 13, Forest Research Division, Bureau of Forestry, Department of Agriculture and Natural Resources, Philippines.

Aguila, M.J., Casate, C., Ferrer, A., Forcade, R.E., Sanchez, R.J., Gonzalez, O.A., Ancizar R.A. and Gonzalez, R. **1989** Uso multiple del suelo con restablecimiento de un bosque poliespecífico periférico. *In*: Memorias de I Congreso Forestal de Cuba y Simposio Internacional sobre técnicas agroforestales, Noviembre 1989. Instituto de Investigaciones Forestales, Cuba.

AIDAB. 1986 A review of the hardwood plantation reforestation subsector in Fiji. *Cited in*: Oliver, W.W. **1992** Plantation forestry in the South Pacific: a compilation and assessment of practices. Field Document 8, USDA Forest Service, Pacific Southwest Research station, Redding, California.

Akanbi, M.O. **1973** The major insect borers of mahogany: a review towards their control. Research Paper (Forest Series) 16, Federal Department of Forest Research, Ministry of Agriculture and Natural Resources, Ibadan, Nigeria.

Alder, D. **1980** Forest volume estimation and yield prediction. FAO Forestry Paper 22(2), Food and Agriculture Organization, Rome, Italy.

Alemán, P.A.M. and García, E. **1974** Climate of Mexico. *In*: R.A. Bryson and F.K. Hare (eds.) *World Survey of Climatology: Climate of North America. Vol. 11.* Elsevier Publishing Co., Amsterdam.

Allan, C.G., Chopra, C.S., Friedhoff, J.F., Gara, R.I., Maggi, M.W., Neogi, A.N., Powell, J.C., Roberts, S.C. and Wilkins, R.M. **1976** The concept of controlled release insecticides and the problem of shoot borers of the Meliaceae. *In*: J.L. Whitmore (ed.) Studies on the shoot borer *Hypsipyla grandella* (Zeller) Lep. Pyralidae, Vol. II, pp. 110-115. IICA Miscellaneous Publications no. 101, CATIE, Turrialba, Costa Rica.

Alrasjid, H. and Mangsud 1973 Natural regeneration trials with mahogany (*Swietenia* spp.) in the Ngraho and Tobo Forest Circles, E. Java. Lembaga Penelitian Hutan 165. 25 pp.

Alston, A.S. 1982 *Timbers of Fiji: properties and potential uses.* Department of Forestry, Suva, Fiji. 183 pp.

Amador, H.C. 1988 Evaluación preliminar del crecimiento de 20 especies maderables en la región de Lloro-Carretera Panamericana, Choco, Colombia. Serie Technico 29, Corporación Nacional de Investigación y Fomento Forestal, CODECHOCO, Bogotá, Colombia.

Andrew, M. 1994 Growth and yield of mahogany (*Swietenia macrophylla* King) plantations in St. Lucia. BSc thesis, University of New Brunswick, Canada.

Anon. 1935 Report on the management and exploitation of the forests of Ceylon. Sessional Paper VII, Ceylon Government Press, Colombo.

Anon. 1944 Forestry in the Leeward Islands. Development and Welfare in the West Indies Bulletin 7, Barbados.

Anon. 1955a Planting instruction for mahogany (*Swietenia macrophylla*), Meliaceae. Occasional Paper 5, Forest Research Division, Bureau of Forestry, Department of Agriculture and Natural Resources, Philippines.

Anon. 1955b Comparative growth of *Swietenia mahagoni* and *Swietenia macrophylla* in the Seychelles. Report no. 14, Department of Agriculture, Seychelles.

Anon. 1956 Germination tests. Report of the Forest Administration of Malaya, Malaysia.

Anon. 1959 *Swietenia macrophylla* King: caractères sylvicoles et méthodes de plantation. *Bois et Forêts des Tropiques* **65**(3): 37-42.

Anon. 1960 Storage of mahogany (*Swietenia macrophylla*) planting stock. Report of the Forestry Department, Fiji.

Anon. 1961 Application of silvicultural methods to some of the forests of the Amazon. Expanded Technical Assistance Program no. 1337. Report to the Government of Brazil, FAO, Rome, Italy.

Anon. 1973 *Afforestation of eroded soils in Java (Indonesia).* State Agricultural University, Wageningen, Netherlands. 64 pp.

Anon. 1975 Yield table for ten industrial wood species. *Cited in*: **Pandey, D. 1983** *Growth and yield of plantation species in the tropics.* W/R0867, FAO, Rome, Italy.

Anon. 1979 Line planting on Kolombangara: choosing the species to match the site. Forest Research Note 12/79, Forestry Division, Solomon Islands.

Anon. 1985 Survey of current techniques in management nurseries. Forest Research Note 2/85, Forestry Division, Solomon Islands.

Anon. 1987a Seed production of seven major plantation species since 1984: implication for future objectives. Forest Research Note 19/3/87, Forestry Division, Solomon Islands.

Anon. 1987b Experimental plan: mahogany and *Securinega* mixture trial. Unpublished research report for the Forestry Division, Solomon Islands.

Anon. 1988a Notes on the use of mixture and nurse crops in the establishment of plantations of high value species. Forest Research Note 30/7/88, Forestry Division, Solomon Islands.

Anon. 1988b Especies florestais nativas e exoticas: comportamento silvicultural no Planalto do Tapajós, Pará. Centro de Pesquisa Agropecuria do Tropico Umido (CPATU), Belém, Brasil.

Appanah, S. and Weinland, G. 1993 Planting quality timber trees in Peninsular Malaysia: a review. Malayan Forest Records 38, Forest Research Institute Malaysia, Kepong, Malaysia.

de Araujo, V.C. 1971 Sobre a germinação do mogno (aguano) *Swietenia macrophylla* King. *Acta Amazonica* **1**(3): 59-69.

Asiddao, F. and Jacalne, D. 1959 Response of mahogany (*Swietenia macrophylla* King) seedlings to underground root pruning. *Philippine Journal of Forestry* **15**(1-4): 13-31.

Ata, A.B. and Ibrahim, Z.B. 1984 Forest plantation development in Peninsular Malaysia: present state of knowledge and research priorites. *In*: L.T. Chim (ed.) *Forest plantation development in Malaysia*. Proceedings from a seminar, July 9-14, Kota Kinabalu, Sabah. Syarikat Art Printing Sdn. Bhd., Sandakan, Sabah.

Aubreville, A. 1953 L'éxperience de l'énrichissement par layons en Côte d'Ivoire. *Bois et Forêts des Tropiques* **29**: 3-9.

de Barros, P.L.C, Queiroz, W.T., Silva, J.N.M., Oliveira, J.N.M.F.deA., Filho, P.P.C., Terezo, E.F.deM., Farias, M.M. and Barros, A.V. 1992 *Natural and artificial reserves of Swietenia macrophylla, King in the Brazilian Amazon - a perspective for conservation*. Belém, Pará, Brazil: Faculdade de Ciências Agropecuaria. 56 pp.

Bascopé, F., Bernadi, A.L. and Lamprecht, H. 1957 Descripciones de árboles forestales no. 1: *Swietenia macrophylla* King. Universidad de los Andes, Mérida, Venezuela.

Bauer, G.P. 1987a *Swietenia macrophylla* and *S. macrophylla* x *S. mahagoni* development and growth: the nursery phase and the establishment phase in line planting in the Caribbean National Forest, Puerto Rico. MSc thesis, College of Environmental Science and Forestry, Faculty of Forestry, State University of New York.

Bauer, G.P. 1987b Reforestation with mahogany (*Swietenia* spp.) in the Caribbean National Forest, Puerto Rico. Paper presented at Seminario-Taller de Cooperación y Manejo de Bosques Tropicales, 3-22 August 1987, Perú.

Bauer, G.P. and Gillespie, A.J.R. 1990 Volume tables for young plantation-grown hybrid mahogany (*Swietenia macrophylla* x *S. mahagoni*) in the Luquillo Experimental Forest of Puerto Rico. Research Paper SO-257, US Department of Agriculture, Forest Service, Southern Forest Experimental Station. 8 pp.

Beard, J.S. 1942 Summary of silvicultural experience with Cedar, *Cedrela mexicana* M. J. Roem. (*C. odorata* L.) in Trinidad. *Caribbean Forester* **3**(3): 91-102.

Becking, J.H. 1928 De djaticultuur op Java. Een vergelijkend onderzoek naar de uitkomsten van verschillende verjongingsmethoden van den djati op Jave. Thesis, Agricultural University, Wageningen, Netherlands.

Berrios, F. and Hidalgo-Salvatierra, O. 1971 Estudios sobre el barrenador *Hypsipyla grandella* Zeller. VIII. Susceptibilidad de la larva a los hongos *Beauvaria bassiana* (Bal.) y *Beauvaria tenella* (Del.). *Turrialba* **21**(4): 451-454.

Betancourt, B.A. 1987 Silvicultura especial de árboles maderables tropicales. Edición Ciencas Téchnicas, La Habonas, Cuba.

Bierne, B. P. 1975 Biological control attempts by introductions against pest insects in the field in Canada. *Canadian Entomology* **107**: 225-236.

Bird, N.M. 1994 Experimental design and background information. Silvicultural Research Paper 1, Forest Planning and Management Division, Ministry of Natural Resources, Belize.

Boone, R.S. and Chudnoff, M. 1970 Variations in wood density of the mahoganies of Mexico and Central America. *Turrialba* **20**(3): 369-371.

Boontawee, B. 1996 Big-leaf mahogany in Thailand. Unpublished report, Division of Silviculture, Forest Research Office, Royal Forest Department, Bangkok.

Bosbouwproefstation, 1949 Herbebossing ban slechte gronden in he Djati-areaal. *Cited in*: **Indonesian Forestry Abstracts 1982** *Dutch literature until about 1960*. Abstract no. 954. Wageningen, Netherlands.

Bramasto, Y. and Harahap, R.M.S. 1989 Seleksi pohon mahoni daun lebar. *Duta Rimba* **113-114**(15): 13-21.

Brienza, S. Jnr. and Yared, J.A.G. 1991 Agroforestry systems as an ecological approach in the Brazilian Amazon development. *Forest Ecology and Management* **45**: 319-323.

Brinkmann, W.L.F. 1970 Optical characteristics of tropical tree leaves: II Mogno (*Swietenia macrophylla* King). Pesquisas Florestais no. 6, Boletím do INPA, Instituto Nacional de Pesquisas da Amazonia, Manaus, Brazil.

Briscoe, C.B., Harris, J.B. and Wyckoff, D. 1963 Variation of specific gravity in plantation grown trees of bigleaf mahogany. *Caribbean Forester* **24**(2): 67-74.

Brokaw, N.V.L., Wilson, R., Whitman, A.A., Hagan, J.M., Bird, N., Martins, P.J., Snook, L.K., Mallory E.P., Darrell, N. White, D. and Losos, E. 1995 Research toward sustainable forestry in the Rio Bravo Conservation and Management Area, Belize. Draft paper, Manomet Observatory for Conservation Sciences, Massachusetts.

Brooks, R.L. 1941 The regeneration of mixed rainforest in Trinidad. *Caribbean Forester* **2**(4): 164-173.

Browder, J.O., Matricardi, E.A.T. and Abdala, W.S. 1996 Is sustainable tropical timber production financially viable? A comparative analysis of mahogany silviculture among small farmers in the Brazilian Amazon. *Ecological Economics* **16**: 147-159.

Building Research Establishment 1972 *A Handbook of Hardwoods*. HMSO, London.

Bulai, S.S. 1993 The properties and potential uses of Fiji mahogany (*Swietenia macrophylla*). Fiji Timbers and their Uses 82, Department of Forestry, Suva, Fiji.

Burger, D. 1924 Voorwararden, waaraan plantensoorten moeten voldoen voor gebruik bij boschbrandbestrijding op Java. *Cited in*: **Indonesian Forestry Abstracts 1982** *Dutch literature until about 1960*. Abstract no. 1590. Wageningen, Netherlands.

Burgos, L.J.A. 1954 Un estudio de la silvicultura de algunas especies forestales en Tingo María, Perú. *Caribbean Forester* **15**(1-2): 14-53.

Busby, R.J.N. 1967 Reforestation in Fiji with large-leaf mahogany. Paper presented at the Ninth Commonwealth Forestry Conference India, 1968. Department of Forestry, Fiji.

Butt, G. and Chiew, S.P. 1982 Guide to site-species matching in Sarawak: selected exotic and native plantations species. Forest Research Report, Forest Department, Sarawak, Malaysia.

Cater, J. C. 1945 The silviculture of *Cedrela mexicana*. *Caribbean Forester* **6**: 89-100.

Chable, A.C. 1967 Reforestation in the Republic of Honduras, Central America. *Ceiba* **13**(2): 1-56.

Chabod, D., in press Big-leaf mahogany and its silviculture in Martinique. Proceedings from the International Conference on Bigleaf Mahogany: biology, genetic resources and management. 22-24 October 1996, San Juan, Puerto Rico.

Chai, F.Y.C. and Kendawang, J.J. 1984 Nursery practice, site evaluation and silvicultural management of plantations in Sarawak. *In*: L.T. Chim (ed.) Forest plantation development in Malaysia. Proceedings from a seminar, July 9-14, Kota Kinabalu, Sabah. Syarikat Art Printing Sdn. Bhd., Sandakan, Sabah.

Chalmers, K., Newton, A.C., Waugh, R., Wilson, J., Powell, W. 1994 Genetic variation in mahoganies (Meliaceae) detected using RAPDs. *Theoretical and Applied Genetics* **89**(4): 504-508.

Chaloner and Fleming 1850 *The Mahogany Tree: its Botanical Characters, Qualities and Uses, with Practical Suggestions for Selecting and Cutting it in the Regions of its Growth*, Chapters III and IV. Rockliff and Son, Liverpool.

Champion, H.G. and Griffith, A.L. 1948 *Manual of General Silviculture for India (Revised Edition)*. Oxford University Press.

Chaplin, G. 1993 Silvicultural manual for the Solomon Islands. ODA Forestry Series no. 1, Overseas Development Administration, London.

Chinte, F.O. 1952 Trial planting of large leaf Mahogany (*Swietenia macrophylla* King). *Caribbean Forester* **13**(2): 75-84.

Chudnoff, M. 1984 Tropical timbers of the world. Agriculture Handbook 607, Forest Service, United States Department of Agriculture.

Chudnoff, M. and Geary, T.F. 1973 On the heritability of wood density in *Swietenia macrophylla*. *Turrialba* **23**(5): 359-362.

CITES. 1997 Proposal to the Convention on International Trade in Endangered Species to include *Swietenia macrophylla* on Appendix II. Government of the Netherlands. Unpublished document.

Cock, M.W. (ed.) 1985 A review of the biological control of pests in the Caribbean and Bermuda up to 1982. CIBC Technical Communication 9, Commonwealth Institute of Biological Control, Trinidad.

Combe, J. and Gewald, N.J. 1979 Guía del campo de los ensayos forestales del CATIE en Turrialba, Costa Rica. Centro Agronómico Tropical de Investigación y Enseñanza, Turrialba, Costa Rica.

Corpuz, V.M. and Cockburn, P.F. 1972 Laporan Penyelidek Hutan 1972. Cawangan Penyelidik Jabatan Hutan, Sabah.

Coster, C. 1934 Tournee rapport no. 18. *Cited in*: **Indonesian Forestry Abstracts 1982** *Dutch literature until about 1960*. Abstract no. 759. Wageningen, Netherlands.

Coster, C. 1935 Licht, ondergroei en wortelconcurrentie. *Cited in*: **Indonesian Forestry Abstracts 1982** *Dutch literature until about 1960*. Abstract no. 386. Wageningen, Netherlands.

Coster, C. 1937 Tournee rapport no. 25. *Cited in*: **Indonesian Forestry Abstracts 1982** *Dutch literature until about 1960*. Abstract no. 786. Wageningen, Netherlands.

Coster, C. 1938 Reisrapport no. 26. *Cited in*: **Indonesian Forestry Abstracts 1982** *Dutch literature until about 1960*. Abstract no. 760. Wageningen, Netherlands.

Coster, C. and Van den Blink, G.C. 1939 Reisrapport no. 1. *Cited in*: **Indonesian Forestry Abstracts 1982** *Dutch literature until about 1960*. Abstract no. 762. Wageningen, Netherlands.

Cottle, G.W. no date Exotic trees in the British Commonwealth: Fiji. Department of Forestry, Fiji.

Cown, D.J., McConchie, D.L. and Kimberley, M.O. 1989 A sawing study of Fijian plantation grown large-leaf mahogany. *Commonwealth Forestry Review* **68**(4): 245-261.

Cree, C.S. 1953 Report of the Forest Department for the year ended 31st December, 1952. British Honduras.

Cree, C.S. 1954 Forest Department Annual Report for 1953. British Honduras.

Cree, C.S. 1955 Forest Department Annual Report for 1954. British Honduras.

Cree, C.S. 1956 Annual Report of the Forest Department for the year 1955. Printing Department, British Honduras.

Cree, C.S. 1957 Annual Report of the Forest Department for the year 1956. Printing Department, British Honduras.

Crocker, W. 1948 *Growth of Plants*. Reinhold Publishing Corp., New York.

Cuevas, G.X., Paraguirre, L.C. and Rodriguez, S.B. 1992 Modelos de crecimiento para una plantación de caoba (*Swietenia macrophylla* King). *Revista Ciencia Forestal en México* **17**(71): 87-102.

Dawkins, H.C. 1963 Crown diameters: their relation to bole diameter in tropical forest trees. *Commonwealth Forestry Review* **42**(4): 318-333.

Dawkins, H.C. 1965 Problems of natural regeneration, plantations and research. Unpublished report for the Forestry Division of the Ministry of Natural Resources, Sierra Leone.

Dayan, M.P., no date Mahogany: *Swietenia macrophylla* King. Source unknown.

DENR. 1992 Forestry Sector Programme. Department of Environment and Natural Resources, Manila, Philippines.

DFR. 1963 Untitled. Technical Note 29, Department of Forest Research, Federal Printing Division, Lagos, Nigeria.

Djapilus, A. 1988 Effects of mahogany (*Swietenia macrophylla*) seeding, pruning and stock type on survival and growth in the field. Buletin Penelitian Hutan 496, Pusat Penelitian dan Pengembangan Hutan, Indonesia.

DoF. 1993 Forestry information 1993. Department of Forests, Ministry of Agriculture, Fisheries and Forests, Fiji.

Dourojeanni, R. M. 1963 El barreno de los brotes (*Hypsipyla grandella*) en cedro y caoba. *Agronomía* **30**(1): 35-43.

Ennion, R.C. 1996 Evaluation of four taungya mahogany increment plots age 36-38 years, Columbia River Forest Reserve. Unpublished report for the Forest Department, Ministry of Natural Resources, Belize.

Entwistle, P.F. 1967 The current situation on shoot, fruit and collar borers of the Meliaceae. Proceedings of the 9th British Commonwealth Forestry Conference, Commonwealth Forestry Institute, Oxford.

Entwistle, P.F. and Evans, H.F. 1983 Viral control. *In*: G.A. Kerkut and L.I. Gilbert (eds.) *Comprehensive Insect Physiology, Biochemistry and Pharmacology*, pp. 347-412. Pergamon Press, Oxford.

ERDB. 1996 Large-leaf mahogany (*Swietenia macrophylla*). Unpublished report, Ecosystems Research and Development Bureau, Department of Environment and Natural Resources, College, Laguna, Philippines. 2 pp.

Evans, J. 1992 *Plantation Forestry in the Tropics (2nd edition)*. Clarendon Press, Oxford.

Ewel, J.J. 1963 Height growth of bigleaf mahogany. *Caribbean Forester* **24**(2): 34-35.

Fattah, H.A. 1992 Mahogany forestry in Indonesia. *In*: Mahogany Workshop: review and implications of CITES, 3-4 February. Tropical Forest Foundation, Washington DC.

Ferron, P. 1981 Pest control by the fungi Beauvaria and Metarhizium. *In*: H.D. Burges (ed.) *Microbial control of pests and plant diseases*, pp. 466-482. Academic Press, London.

Flachsenberg, H. 1994 Regeneración natural en patios de concentratión de trocería y el incremento medio anual de la caoba. Unpublished report for the Acuerdo México Alemania, Chetumal, Mexico.

Flinta, C.M. 1960 Prácticas de plantación forestal en América Latina. Cuadernos de Fomento Forestal no. 15, FAO, Rome, Italy.

Forest Department, no date Research Programme 1965-1969 and 1965 Supplement. Forest Department, Sabah.

Forest Division, 1987 Research Annual Report. Forest Division, Ministry of Natural Resources, Solomon Islands.

Forest Division, 1990 Research Annual Report. Forest Division, Ministry of Natural Resources, Solomon Islands.

Forestry Division, 1979 The performance of forest plantation species in Dominica 1971-8. Forestry Division, Ministry of Agriculture, Dominica.

Foury, P. 1956 Comparison des méthodes d'enrichissement utilisées en forêt dense équatoriale. *Bois et Forêts des Tropiques* **47**: 15-25.

Francis, J.K. 1991 West Indies Mahogany. Forest Research Paper SO-ITF-SM-46, International Institute of Tropical Forestry, USDA, Puerto Rico.

Francis, J.K. 1995 Forest plantations in Puerto Rico. *In*: A.C. Lugo and C. Lowe (eds.) *Tropical Forests: Management and Ecology*. Springer-Verlag, New York.

Francis, J.K. 1996 Species for Forest Plantations in Puerto Rico. Draft paper, International Institute of Tropical Forestry, USDA, Puerto Rico.

Francis, J.K., in press Mahogany planting and research in Puerto Rico and the US Virgin Islands. Proceedings from the International Conference on Bigleaf Mahogany: biology, genetic resources and management. 22-24 October 1996, San Juan, Puerto Rico.

Frederick, R. 1994 An interim evaluation of line planted mahogany at Grand Bras Estate, Grenada. Unpublished report for the International Institute of Tropical Forestry, USDA, Puerto Rico.

Frith, A.C. 1958 Annual Report of the Forest Department for the year 1957. Printing Department, British Honduras.

Frith, A.C. 1959 Annual Report of the Forest Department for the year 1958. Printing Department, British Honduras.

Frith, A.C. 1960 Annual Report of the Forest Department for the year 1959. British Honduras.

Frith, A.C. 1961 Annual Report of the Forest Department for the year 1960. The Government Printer, British Honduras.

Frith, A.C. 1962 Annual Report of the Forest Department for the year 1961. Printing Department, British Honduras.

Frith, A.C. 1963 Annual Report of the Forest Department for the year 1962. British Honduras.

Frith, A.C. 1964 Annual Report of the Forest Department for the year 1963. Printing Department, British Honduras.

Funes, L.S., San Martín, V.A. and Henriquez, N.A. 1983 Tablas de volumén local y general y algunas relaciones dasométricas para *Swietenia macrophylla* (caoba). Tesis, Escuela Nacional de Ciencias Forestales, Siguatepeque, Honduras.

García, X.C., Negreros, P.C. and Rodriguez, B.S. 1993 Regeneración natural de caoba (*Swietenia macrophylla* King) bajo diferentes densidades de dosel. *Revista Ciencia Forestal en México* **18**(74): 25-43.

Geary, T.F. 1969 Adaptability of Mexican and Central American provenances of *Swietenia* in Puerto Rico and St Croix. Second World Consultation on Forest Tree Breeding, Washington, 7-16th August, FAO-IUFRO.

Geary, T.F., Barres, H. and Ybarra-Coronado, R. 1973 Seed source variation in Puerto Rico and Virgin Islands grown mahoganies. Forest Research Paper ITF-17, Institute of Tropical Forestry, Puerto Rico.

Geary, T.F., Nobles, R.W. and Briscoe, C.B. 1972 Hybrid mahogany recommended for planting in the Virgin Islands. Forest Research Paper ITF-15, Institute of Tropical Forestry, Puerto Rico.

Geigel, F.B. 1977 Materia orgánica y nutrientes devueltos al suelo mediante la hojarasca de diversas especies forestales. *Baracoa* **7**(3/4): 15-38.

Gerhardt, K. 1994 Seedling development of four tree species in secondary tropical dry forest in Guanacaste, Costa Rica. PhD thesis, Department of Ecological Botany, Uppsala University, Sweden.

Gibson, I.A.S. 1975 *Diseases of Forest Trees Widely Planted as Exotics in the Tropics and Southern Hemisphere. Part 1: Important Members of the Myrtaceae, Leguminosae, Verbenaceae and Meliaceae.* Commonwealth Forestry Institute, University of Oxford.

Glogiewicz, J.S. 1986 Performance of Mexican, Central American and West Indian provenances of *Swietenia* grown in Puerto Rico. MSc thesis, College of Environmental Science and Forestry, Faculty of Forestry, Syracuse, New York.

Golfari, L. and Caser, R.L. 1977 Zoneamento Ecológico da Região Nordeste Pará: experientação florestal. Serie Técnica no. 10, Centro de Pesquisa Florestal da Região do Cerrado, Belo Horizonte, Brasil.

Gonggrijp, L. 1940 Maandverslag no. 7 en 8 van den Opperhoutvester bij het Boschbouwproefstation over Juli/Augustus 1940. *Cited in*: **Indonesian Forestry Abstracts 1982** *Dutch literature until about 1960.* Abstract no. 61. Wageningen, Netherlands.

Granert, W.G. and Cadampog, Z. 1980 Effects of a leguminous nurse tree, Giant Ipil-ipil (*Leucaena leucocephala* (L.) Benth.) on the growth rate of teak (*Tectona grandis* L.) and mahogany (*Swietenia macrophylla* King) in Talamban, Cebu City. *The Philippine Scientist* **17**: 137-145.

Grey, G.A.L. 1990 Silvicultural treatments for the increased production of *Swietenia macrophylla* (King) sawlogs via the control of *Hypsipyla grandella* (Zeller). BSc thesis, University of New Brunswick, Canada.

Grijpma, P. 1972 Studies on the shootborer *Hypsipyla grandella* (Zeller) (Lepidoptera, Pyralidae). X. Observations on the egg parasite *Trichogramma semifumatum* (Perkins) (Hym., Trichogrammatidae). *Turrialba* **22**(4): 398-402.

Grijpma, P. 1973 Studies on the shootborer *Hypsipyla grandella* (Zeller) (Lep., Pyralidae). XVII. Records of two parasites new to Costa Rica. *Turrialba* **23**(2): 235-236.

Grijpma, P. 1974 *Contributions to an Integrated Control Programme of* Hypsipyla grandella *(Zeller) in Costa Rica.* Wageningen, Netherlands.
Grijpma, P. 1976 Resistance of Meliaceae against the shoot borer *Hypsipyla* with particular reference to *Toona ciliata* M. J. Roem. var *australis* (F. v. Muell.) C.DC. *In*: J. Burley and B.T. Styles (eds.) *Tropical Trees: Variation, Breeding and Conservation*, pp. 69-78. Linnean Society, London.
Grijpma, P. and Gara, R.I. 1970 Studies on the shootborer *Hypsipyla grandella* (Zeller). I. Host selection behaviour. *Turrialba* 20(2): 233-240.
Grijpma, P. and Styles, B.T. 1973 *Bibliografia selectiva sobre Meliaceas.* Turrialba, Costa Rica, IICA-CTEI.
Gullison, R.E. 1995 Conservation of tropical forests through the sustainable production of forest products: the case of mahogany (*Swietenia macrophylla* King) in the Chimanes Forest, Beni, Bolivia. PhD thesis, Princeton University, USA.
Gullison, R.E. in press Can mahogany be conserved through sustainable use? *In*: E.J. Milner Gulland and R. Mace (eds.) *The Conservation and Use of Biological Resources.* Blackwell, Oxford.
Gullison, R.E., Panfil, S.N., Strouse, J.J. and Hubbell, S.P. 1996 Ecology and management of mahogany (*Swietenia macrophylla* King) in the Chimanes Forest, Beni, Bolivia. *Botanical Journal of the Linnean Society*, 122(1): 9-34.
de Guzman, E.D. and Eusebio, E.C. 1975 Root rot of *Swietenia macrophylla* King seedlings. *Pterocarpus* 1: 64-65.
Hart, H.M.J. 1925 Wortelschimmel in djaticulturen. Cited *in*: **Indonesian Forestry Abstracts 1982** *Dutch literature until about 1960.* Abstract no. 1531. Wageningen, Netherlands.
Haslett, A.N. 1986 Properties and uses of the timbers of Western Samoa: plantation grown exotic hardwoods. Ministry of Foreign Affairs, Wellington, New Zealand.
Hauxwell, C., Mayhew, J.E. and Newton, A.C., in press Silvicultural control of the shoot borer. Proceedings from the International Workshop on *Hypsipyla* Shoot-Borers in the Meliaceae. 20-23 August 1996, Kandy, Sri Lanka.
Hazlett, D.L. and Montesinos, J.L. 1980 El crecimiento de 27 especies maderables en plantaciones de Lancetilla. Artículo Científico 1, Escuela Nacional de Ciencias Forestales, Siguatepeque, Honduras.
Helgason, T., Russell, S.J., Monro, A.K. and Vogel, J.C. 1996 What is mahogany? The importance of a taxonomic framework for conservation. *Botanical Journal of the Linnean Society* 122: 47-59.
Hepburn, A.J. 1969 Annual Report of the Research Branch of the Forest Department. State of Sabah, Malaysia.
Hérard, F., Keller, M.A., Lewis, W.J. and Tumlinson, J.H. 1988 Beneficial arthropod behaviour mediated by airborne semiochemicals. III. Influence of age and experience on flight chamber responses of *Microplitis demolitor* Wilkinson. *Journal of Chemical Ecology* 14(7): 1583-1596.
Hidalgo-Salvatierra, O. and Berrios, F. 1972 Studies on the shootborer *Hypsipyla grandella* (Zeller) (Lep., Pyralidae) XII. Determination of the LD50 of *Metarhizum anisopliae* (Metchnikoff) Sorokin spores on fifth instar larvae. *Turrialba* 22(4): 435-438.
Hidalgo-Salvatierra, O. and Palm, J.D. 1972 Studies on the shootborer *Hypsipyla grandella* (Zeller) (Lep., Pyralidae) XIV. Susceptibility of first instar larvae to *Bacillus thuringiensis. Turrialba* 22(4): 467-478.

Hokkanen, H.M.T. and Pimentel, D. 1989 New associations in biological control: theory and practice. *Canadian Entomology* **121**: 829-840.

Holdridge, L.R. 1943 Comments on the silviculture of *Cedrela*. *Caribbean Forester* **4**: 77-80.

Holdridge, L.R. and Marrero, J. 1940 Preliminary notes on the silviculture of the big-leaf mahogany. *Caribbean Forester* **2**(1): 20-23.

Holdridge, L.R. and Poveda, L.J. 1975 Árboles de Costa Rica, Vol. 1. Centro de Ciencias Tropicales, San José, Costa Rica.

Holl, K., Daily, G. and Ehrlich, P.R. 1990 Integrated pest management in Latin America. *Environmental Conservation* **17**(4): 341-350.

Howard, A.F., Rice, R.E. and Gullison, R.E. 1996 Simulated financial returns and selected environmental impacts from four alternative silvicultural prescriptions applied in the neotropics: a case study of the Chimanes Forest, Bolivia. *Forest Ecology and Management* **89**: 43-57.

Howard, F.W., Verkade, S.D. and DeFilippis, J.D. 1988 Propagation of West Indies, Honduran and hybrid mahoganies by cuttings, compared with seed propagation. *Proceedings of the Florida State Horticultural Society* **101**: 296-298.

Hoy, H.E. 1946 Mahogany industry of Peru. *Economic Geography* **22**(1): 1-13.

Huguet, L. and Marie, E., 1951 Les plantations d'Acajou d'Amérique des Antilles Francaises. *Bois et Forêts des Tropiques* **17**(1): 12-25.

Hummel, C. 1925 Report on the Forests of British Honduras. Crown Agents for the Colonies.

Hutchinson, I.D. 1991 Diagnostic sampling to orient silviculture and management in natural tropical forest. *Commonwealth Forestry Review* **71**(2): 113-132.

Ikeda, T., Taketani, A., Matumoto, K. and Yokota, A., in press Management of the shoot-borer: case studies in Peru and Indonesia. Proceedings from the International Conference on Bigleaf Mahogany: biology, genetic resources and management. 22-24 October 1996, San Juan, Puerto Rico.

de Irmay, H. 1948 La Caoba en Bolivia. Boletín Forestal no. 1, Publicaciones de la Facultad de Ciencias Agronómicas, Universidad Mayor de San Simón. Imprenta Universitaria, Cochabamba, Bolivia.

Jacalne, D.V., Meimban, J.R. and Tadle, J.F. 1959 A study on the stump planting of mahogany (*Swietenia macrophylla* King). *Philippine Journal of Forestry* **13**(1/2): 63-80.

James, R.A. 1990 No title. Unpublished letter to J.K. Francis, International Institute of Tropical Forestry, USDA, Puerto Rico.

Japing, H.W. and Seng, O.D. 1936 Cultuurproeven met wildhoutsoorten in Gadoengan VI. Cited in: **Indonesian Forestry Abstracts 1982** *Dutch literature until about 1960*. Abstract no. 807. Wageningen, Netherlands.

JICA. 1982 Report of site assessment on plantation of hard-wood in Viti Levu, Fiji. Japan International Cooperation Agency report for the Department of Forestry, Fiji.

Jones, N. 1976 Kolombangara reforestation project feasibility study: tree breeding report for CDC. Unpublished report for the Commonwealth Development Corporation, London. 40 pp.

Kalshoven, L.G.E. 1926 Primaire aantasting van houtige gewassen door *Xyleborus*-soorten. Cited in: **Indonesian Forestry Abstracts 1982** *Dutch literature until about 1960*. Abstract no. 1675. Wageningen, Netherlands.

Kamath, M.K., Senivasa, E. and Bola, I. 1993 Impact of termites on mahogany (*Swietenia macrophylla*) plantations in Fiji. IUFRO Proceedings 93.1, Kerala Forestry Research Institute, Peechi, Kerala, India. 9 pp.

Kamath, M.K., Senivasa, E. and Raravula, T.G. 1995 Termite damage in forest trees in Fiji. *Fiji Agricultural Journal* **51**(1): 9-19.

Kareira, P. 1983 Influence of vegetation texture on herbivore populations: resource concentration and herbivore movement. *In*: R.F. Denno and M.S. McClure (eds.) *Variable plants and herbivores in natural and managed systems*, pp. 259-289. Academic Press, New York.

Keay, R.W.J. 1996 Introduction: the future of the genus *Swietenia* in its native forests. *Botanical Journal of the Linnean Society* **122**(1): 3-7.

Keys, M.G. and Nicholson, D.I. 1982 Underplanting of silviculturally treated rainforest in North Queensland. Unpublished report, Department of Forestry, Queensland. 22 pp.

Kinloch, J.B. 1933 Report of the Forest Trust for the Biennial Period ending March 31st 1933. The Government Printer, British Honduras.

Kinloch, J.B. 1938 Report of the Forest Department for the Year 1937. The Government Printer, British Honduras.

Koul, O. and Isman, M. B. 1992 Toxicity of the limonoid allelochemical cedrelone to noctuid larvae. *Entomologia Experimentalis et Applicata* **64**: 281-287.

Kriek, W. 1970 Report to the government of Uganda on performance of indigenous and exotic trees in species trials. TA2826, FAO, Rome, Italy.

Krohn, T.J. 1981 One-year results of species trials on Guam. *Tree Planters Notes* **32**(2): 30-34.

Lamb, A.F.A. 1945 Abridged Report of the Forest Department for the Year 1944. The Government Printer, British Honduras.

Lamb, A.F.A. 1946 Report of the Forest Department for the Year 1945. British Honduras.

Lamb, A.F.A. 1947 Report of the Forest Department for the Year 1946. British Honduras.

Lamb, A.F.A. 1948 Report of the Forest Department for the Year 1947. British Honduras.

Lamb, A.F.A. 1949 Report of the Forest Department for the Year ended 31st December, 1948. British Honduras.

Lamb, A.F.A. 1955 Forestry on private estates. *Journal of the Agricultural Society of Trinidad and Tobago* **55**(2): 169-183.

Lamb, A.F.A. 1969 Artificial regeneration within the humid lowland tropical forest. *Commonwealth Forestry Review* **48**(1): 41-53.

Lamb, F.B. 1960 An approach to mahogany tree improvement. *Caribbean Forester* **21**(1/2): 12-20.

Lamb, F.B. 1963 On further defining mahogany. *Economic Botany* **17**: 217-232.

Lamb, F.B. 1966 *Mahogany of Tropical America: its Ecology and Management*. University of Michigan Press, Ann Arbor.

Lamprecht, H. 1989 *Silviculture in the Tropics*. Deutsche Gesellschaft für Technische Zusammenarbeit (GTZ), Germany.

Langley, R.G. 1963 The reaction of British Honduras indigenous hardwood species to hurricaine Hattie. Unpublished report for the Forest Department, British Honduras.

Lanuza, R.L. 1991 Performance of potted large-leaf mahogany (*Swietenia macrophylla* King) seedlings at different nitrogen regimes under nursery condition. *Ecosystems Research Digest* **1**(1): 12-18.

Lapis, E.B. 1995 Mahogany shoot borer. Unpublished report for the Center for Forest Pest Management and Research, Department of Environment and Natural Resources, College, Laguna, Philippines. 5 pp.

Leakey, R.R.B., Last, F.T. and Longman, K.A. 1982 Domestication of forest trees: a process to secure the productivity and future diversity of tropical ecosystems. *Commonwealth Forestry Review* **61**: 33-42.

Leakey, R.R.B., Mesén, J.F., Tchoundjeu, Z., Longman, K.A., Dick, J.McP., Newton, A., Matin, A., Grace, J., Munro, R.C. and Muthoka, P.N. 1990 Low-technology techniques for the vegetative propagation of tropical trees. *Commonwealth Forestry Review* **69**(3): 247-257.

Leakey, R.R.B. and Newton, A.C. (eds.) 1994 *Domestication of tropical trees for timber and non-timber products*. MAB Digest 17, UNESCO, Paris.

Leakey, R.R.B., Newton, A.C. and Dick, J.McP. 1994 Capture of genetic variation by vegetative propagation: processes determining success. *In*: R.R.B. Leakey and A.C. Newton (eds.) *Tropical trees: the potential for domestication and the rebuilding of forest resources*, pp. 72-83. HMSO, London.

Lee, H. 1968 Preliminary report on the juvenile characters and heterosis of the hybrids between *Swietenia mahagoni* x *S. macrophylla*. *Taiwania* **14**: 43-52.

Lee, S.K. and Rao, A.N. 1988 Plantlet production of *Swietenia macrophylla* King through tissue culture. *Gardens Bulletin, Singapore* **41**: 11-18.

Leitch, J. 1991 Mahogany site index curves. Technical Report 1991/2, Management Services Division, Department of Forestry, Fiji.

Leslie, A. 1994 A compilation of results from forestry trials established on Espiritu Santo, Vanuatu. Technical Booklet no. 3, Santo Industrial Forest Plantation Project, Vanuatu.

Lindo, L.S. 1968 The effect of hurricanes on the forests of British Honduras. Paper presented at the ninth Commonwealth forestry conference, India 1968. Government Printing Department Belize, British Honduras.

Little, E.L., Wadsworth, F.H. and Marrero, J. 1967 Árboles comunes de Puerto Rico y las Islas Vírgenes. Editorial UPR, Río Piedras, Puerto Rico. 806 pp.

Liu, C.-P. 1970 The genetic improvement of Honduras mahogany. I. Studies on natural variation and individual selection. *Quarterly Journal of Chinese Forestry* Taipei **3**(3): 41-56.

Llera, Z.M. and Melandez, N.F. 1989 Evaluación de especies forestales tropicales como alternativa para la sustitución del arbol de sombra mote (*Erythrina* spp.) en el cultivo del cacao. *In*: N.L. Linares (ed.) Simposio agroforestal en México, Tomo 1. Facultad de Ciencias Forestales, Universidad Autonoma de Nuevo Leon, Mexico.

Lobato, R.C. 1969 Ensaio de localização das faixas com maiores dimensões e maior peso de sementes no fruto maduro inteiro de *Swietenia macrophylla* King. *Ciencia e cultura* **21**(2): 440.

Lugo, A.E. and Liegel, L.H. 1987 Comparison of plantations and natural forests in Puerto Rico. *In*: A.E. Lugo, J.J. Ewel, S.B. Hecht, P.G. Murphy, C. Padoch, M.C. Schmink and D. Stone (eds.) *People and the tropical forest*. United States Man and the Biosphere Program, Tropical and Sub-tropical Forests Directorate, United States Department of State. US Government Printing Office, Washington DC.

Lushington, P.M. 1921 Ceylon Forests in the Year 1921 and their General Administration. Government Printer, Ceylon.

Lyhr, K.P. 1992 Mahogany: silviculture and use of American mahogany (*Swietenia* spp.). Report for the Unit of Forestry, Department of Economics and Natural Resources, The Royal Veterinary and Agricultural University, Copenhagen.

McConchie, D.L., McKinley, R.B., Kimberly, M.O. and Cown, D.J. 1989 Fiji mahogany sawing trials. Part II: site comparisons and validation of log grade relativities. Wood Technology Division Contract Report no. 1349, Wood Technology Division, Forest Research Institute, Rotorua, Fiji.

McNeil, W.M. 1935 Notes on exotic forest trees in Ceylon. Ceylon Government Press, Colombo.

Maruyama, E. and Carrera, F. 1989 Girdling of trees supplemented with applications of Glyphosate in a line planting in the Peruvian Amazonic Zone. *Journal of the Japanese Forestry Society* **71**(9): 369-373.

Maruyama, E., Ishii, K., Satio, A. and Migita, K. 1989 Screening of suitable sterilization of explants and proper media for tissue culture of eleven tree species of Peru-Amazon forest. *Journal of Agricultural Science* **33**(4): 252-261.

Maldonado, E.D. and Boone, R.S. 1968 Shaping and planing characteristics of plantation-grown mahogany and teak. Forest Service Research Paper ITF-7, Institute of Tropical Forestry, Río Piedras, Puerto Rico.

Malone, K. 1963 Notes on the word mahogany. *Economic Botanist* **19**(3): 286-292.

Marcano, A.R. 1963 Discusión de algunas experiencias relativas a ensayos de crecimiento con las especies cedro (*Cedrela mexicana* Roem) y Caoba (*Swietenia macrophylla* King). *Boletín del Instituto Forestal Latino Americano de Investigación y Capacitación* **13**: 38-50.

Marie, E. 1949 Notes on reforestation with *Swietenia macrophylla* King in Martinique. *Caribbean Forester* **10**(3): 206-222.

Marin, E.T. 1963 Preliminary study on the growth and survival of underground root-pruned mahogany (*Swietenia macrophylla* King) and ipil (*Intsia bijuga*) seedlings in field planting. Occasional Paper 11, Forest Research Division, Bureau of Forestry, Department of Agriculture and Natural Resources, Philippines.

Marrero, J. 1942 Study of grades of broadleaved mahogany planting stock. *Caribbean Forester* **3**(20): 79-87.

Marrero, J. 1949 Tree seed data from Puerto Rico. *Caribbean Forester* **10**(1): 11-30.

Marrero, J. 1950 Results of forest planting in the insular forests of Puerto Rico. *Caribbean Forester* **11**(1): 107-147.

Marshall, R.C. 1939 *Silviculture of the Trees of Trinidad and Tobago*. Oxford University Press.

Maruyama, E. and Carrera, F. 1989 Girdling of trees supplemented with applications of Glyphosate in a line planting in the Peruvian Amazonic Zone. *Journal of the Japanese Forestry Society* **71**(9): 369-373.

Matos, J., Sousa, S., Wandelli, E., Perin, R., Arcoverde, M. and Fernandes, E., in press Performance of big-leaf mahogany in agroforestry systems in the western Amazon region of Brazil. Proceedings from the International Conference on Bigleaf Mahogany: biology, genetic resources and management. 22-24 October 1996, San Juan, Puerto Rico.

Matthews, J.D. 1992 *Silvicultural Systems*. Clarendon Press, Oxford.

Mayhew, J.E., in press a Site index curves for *Swietenia macrophylla. Sri Lanka Forester.*

Mayhew, J.E., in press b Volume tables for *Swietenia macrophylla. Sri Lanka Forester.*

Mayhew, J.E., Andrew, M., Sandom, J.H., Thayaparan, S. and Newton, A.C., in press Silvicultural systems for mahogany (*Swietenia macrophylla*) plantations. Proceedings from the International Conference on Bigleaf Mahogany: biology, genetic resources and management. 22-24 October 1996, San Juan, Puerto Rico.

Mijers, W.N. 1941 Herbebossching op Madoera. *Cited in*: **Indonesian Forestry Abstracts 1982** *Dutch literature until about 1960*. Abstract no. 5468. Wageningen, Netherlands.

Miller, N.G. 1990 The genera of Meliaceae in the south-eastern United States. *Journal of the Arnold Arboretum* **71**: 453-486.

Miller, W.A. 1941 Mahogany logging in British Honduras. *Caribbean Forester* **2**(2): 67-72.

Mondala, C.A. 1977 Depth and position of sowing large-leaf Mahogany seeds. *Sylvatrop* **2**(2):131-137.

Montesinos, J.L., Navarro, C.A. and Romero Puerto, J.O. 1985 Resultados de 31 especies forestales en ensayos para plantaciones de Lancetilla. Jardín Botánico de Lancetilla, Honduras.

Morgan, F.D. and Suratmo, F.G. 1976 Host preferences of *Hypsipyla robusta* (Moore) in West Java. *Australian Forestry* **39**(2): 103-112.

MSD. 1996 Untitled. Information from plantation data bank, Management Services Division, Department of Forestry, Suva, Fiji.

Muttiah, S. 1965 A comparison of three repeated inventories of Sundapola mixed selection working circle and future management. *Ceylon Forester* **7**(1): 3-35.

Myers, J. G. 1935 Second report on an investigation into the biological control of the West Indian insect pests. *Bulletin of Entomological Research* **26**: 181-252.

Nasi, R. and Monteuuis, O. 1993 Un nouveau programme de recherches au Sabah (2^e partie): les arbres. *Bois et Forêts des Tropiques* **235**(1): 25-28.

Nastor, M. 1957 Height growth and mortality of seven year old large leaf mahogany (*Swietenia macrophylla* King) saplings in Santa subsidiary nursery, Caniaw Reforestation Project, Santa, Ilocos Sur. Occasional Paper 27, Forest Research Division, Bureau of Forestry, Department of Agriculture and Natural Resources, Philippines.

Nazif, M. 1992 Efficacy test of some herbicides to control weeds under mahogany (*Swietenia macrophylla* King). Buletin Penelitian Hutan 547, Pusat Penelitian dan Pengembangan Hutan, Indonesia.

Negreros, P. 1991 Ecology and management of mahogany (*Swietenia macrophylla* King) regeneration in Quintana Roo, Mexico. PhD thesis, Department of Forestry, Iowa State University, Ames.

Negreros, P., in press Enrichment planting and the sustainable harvest of mahogany in Quintana Roo, Mexico. Proceedings from the International Conference on Bigleaf Mahogany: biology, genetic resources and management. 22-24 October 1996, San Juan, Puerto Rico.

Neil, P.E. 1986 *Swietenia macrophylla* (mahogany) in Vanuatu. Forest Research Report no. 4/86, Vanuatu Forest Service, Port Villa, Vanuatu.

Neil, P.E. and Barrance, A.J. 1987 Cyclone damage in Vanuatu. *Commonwealth Forestry Review* **66**(3): 255-264.

Nelson Smith, J.H. 1942 The formation and management of mahogany plantations at Silk Grass Forest Reserve. *Caribbean Forester* **3**(2): 75-78.
Nelson, R.E. and Schubert, T.H. 1976 Adaptability of selected tree species planted in Hawaii forests. Resource Bulletin no. 14, Pacific Southwest Forest and Range Experiment Station, USDA Forest Service, Honolulu, Hawaii.
Newton, A.C. 1993 Prospects of growing mahogany in plantations. *In*: The Proceedings of the Second Pan American Furniture Manufacturers' Symposium on Tropical Hardwoods, November 3-5 1992. Center for Environmental Study, USA.
Newton, A.C., Baker, P., Ramnarine, S., Mesén, J.F. and Leakey, R.R.B. 1993a The mahogany shoot borer: prospects for control. *Forest Ecology and Management* **57**: 301-328.
Newton, A.C., Cornelius, J.P., Baker, P., Gillies, A.C.M., Hernández, M., Ramnarine, S., Mesén, J.F. and Watt, A.D. 1996 Mahogany as a genetic resource. *Botanical Journal of the Linnean Society* **122**: 61-73.
Newton, A.C., Cornelius, J.P., Mesén, J.F., Corea, E.A. and Watt, A.D., in press Variation in attack by the mahogany shoot borer, *Hypsipyla grandella* (Lepidoptera: Pyralidae), in relation to host growth and phenology. *Bulletin of Entomological Research*.
Newton, A.C., Cornelius, J.P., Mesén, J.F. and Leakey, R.R.B. 1995 Genetic variation in apical dominance of *Cedrela odorata* seedlings in response to decapitation. *Silvae Genetica* **44**: 146-150.
Newton, A.C., Leakey, R.R.B. and Mesén, J.F. 1993b Genetic variation in mahoganies: Its importance, utilization and conservation. *Biodiversity and Conservation* **2**: 114-126.
Newton, A.C., Leakey, R.R.B., Powell, K., Chalmers, K., Waugh, R., Tchoundjeu, Z., Mathias, P.J., Alderson, P.G., Mesén, J.F., Baker, P. and Ramnarine, S. 1994 Domestication of mahoganies. In: R.R.B. Leakey and A.C. Newton (eds.) *Tropical Trees: the potential for domestication and the rebuilding of forest resources*, pp. 256-266. HMSO, London.
Niebroek, H. 1940 Jaarverslag van de beheerseenheid Bojonegoro 1939. *Cited in*: **Indonesian Forestry Abstracts 1982** *Dutch literature until about 1960*. Abstract no. 4502. Wageningen, Netherlands.
Nobles, R.W. and Briscoe, C.B. 1966 Root pruning of mahogany nursing stock. Forest Service Research Note 7, Institute of Tropical Forestry, Río Piedras, Puerto Rico.
Noltee, A.C. 1926 *Swietenia mahagoni* and *Swietenia macrophylla*. *Cited in*: **Indonesian Forestry Abstracts 1982** *Dutch literature until about 1960*. Abstract no. 735. Wageningen, Netherlands.
Oliphant, J.N. 1925 Report of the Forest Trust. Clarion Press, British Honduras.
Oliphant, J.N. 1926 Report of the Forest Trust. Clarion Press, British Honduras.
Oliphant, J.N. 1928 Report of the Forest Trust. Clarion Press, British Honduras.
Oliver, W.W. 1992 Plantation forestry in the South Pacific: a compilation and assessment of practices. Field Document 8, USDA Forest Service, Pacific Southwest Research station, Redding, California.
Ompad, L.S. 1947 Propagation of large-leaf mahogany (*Swietenia macrophylla* King) by cutting. BSc thesis, College of Forestry, Los Baños, University of the Philippines.
Orden, T. 1956 Preparation of local volume table for large-leaf mahogany (*Swietenia macrophylla* King) and its application. *Philippine Journal of Forestry* **12**(3/4): 117-132.

Pacheco, A.A.R. and Chavira, J.M.B. 1979 Desarollo de caoba (*Swietenia macrophylla* King) en diferentes tipos de suelos. *Revista Ciencia Forestal en México* **22**(4): 45-64.

Palmer, J.R. 1994 Designing commercially promising tropical timber species. *In*: R.R.B. Leakey and A.C. Newton (eds.) *Tropical Trees: the potential for domestication and the rebuilding of forest resources*, pp. 16-24. HMSO, London.

Pandey, D. 1983 Growth and yield of plantation species in the tropics. W/R0867, FAO, Rome, Italy.

Pandey, D. 1987 Yield model of plantations in the tropics. *Unasylva* **157/158**(39): 74-75.

Parraguirre, C.L. and Camacho, M.F. 1992 Velocidad de germinación de veintiuno especies forestales tropicales. *Revista Ciencia Forestal en México* **17**(72): 3-26.

Parraguirre, C.L. and Cetz, R.C. 1989 Determinación de la madurez fisiológica de semillas de caoba (*Swietenia macrophylla* King). Congreso Forestal Mexicano (Tomo II), Mexico.

Patiño V.F. 1997 Genetic resources of *Swietenia* and *Cedrela* in the neotropics: proposals for coordinated action. FAO, Rome, Italy.

Pawsey, R.G. 1970 Forest diseases in Trinidad and Tobago with some observations in Jamaica. *Commonwealth Forestry Review* **49**(1): 64-77.

Pedroso, L.M. and Pereira, A.P. 1971 Informacões preliminares sobre a silvicultura de 38 especies florestais da estação experimental de Curua-Una. Assessoria de Programação e Coordenação Divisão de Documentação, Belém, Brasil.

Pennington, T.D. 1981 *A Monograph of Neotropical Meliaceae*. The New York Botanical Garden, New York.

Pennington, T.D. and Styles, B.T. 1975 A generic monograph of the Meliceae. *Blumea* **22**(3): 419-540.

Perera, S.P. 1955 *Swietenia macrophylla* and its propagation by striplings in Ceylon. *Ceylon Forester* **2**(2): 75-79.

PfB. 1996 Rio Bravo Conservation and Management Area Management Plan (3rd edition). Programme for Belize, Belize City.

PfB. 1997 Planning Guidelines for Timber Extraction on the Rio Bravo Conservation and Management Area. Unpublished report by Programme for Belize, Belize City.

Plumptre, A.J. 1995 The importance of 'seed trees' for the natural regeneration of selectively logged tropical forest. *Commonwealth Forestry Review* **74**(3): 253-258.

Pollard, J.F. 1983 Planting of mahogany in mixture. Unpublished report for the Forestry Division, Solomon Islands.

Ponce, S.S. 1933 Mahogany as a reforestation crop. *The Makiling Echo* **12**(1): 13-33.

Poras, J.M. and Luyano, G.B. 1974 Es posible mediante el sistema taungya aumentar la productividad de los bosques tropicales en México? Boletín technico no. 39, Instituto Nacional de Investigaciones Forestales, México.

Portig, W.H. 1976 Climate of Central America. *In*: W. Schwerdtfeger (ed.) *World Survey of Climatology: Climate of Central and South America. Vol. 12*. Elsevier Publishing Co., Amsterdam.

Pottier, D. 1984 Running cattle under trees: an experiment in agroforestry. *Unasylva* **36**(143): 23-27.

Poupon, J. 1982 Carte forestière de la Martinique: notice explicative. Direction Régionale de l'Office National des Forêts, Département de la Martinique.

Puerto, J.O.R. 1983 Crecimiento en dos plantaciones de caoba (*Swietenia macrophylla*) y su regeneración natural vista en Lancetilla. BSc thesis, Universidad Nacional Autónoma de Honduras.

Quintana, S.B. 1986 *Leucena leucocephala* (Lam.) de Wit. as a nurse species for immediate reforestation of watershed areas in the Philippines. *Leucena Research Reports* **7**: 102.

Quirino, W.F., Nakamura, R.M., Lisboa, C.D.J. and de Brito, C.T. 1982 Resistencia da madeira de quatro especies florestais ao ataque do fungo *Trichoderma viride*. Laboratorio de Produtos Florestais, Departamento de Economía Florestal, Instituto Brasileiro de Desevolvimento Florestal, Ministerio da Agricultura, Brasil.

Ramirez Sanchez, J. 1964 Investigación preliminar sobre biología, ecología y control de *Hypsipyla grandella* Zeller. *Bolletín del Instituto Forestal Latino Americano de Investigación y Capacitación* **16**: 54-77.

Ramirez Sanchez, J. 1966 Apuntes sobre contról de *Hypsipyla grandella* Zeller con insecticidas. *Boletín Instituto Forestal de Latino America de Investigación y Capacitación* **22**: 33-37.

Ramos, G. and Grace, J. 1990 The effects of shade on the gas exchange of seedlings of four tropical trees from Mexico. *Functional Ecology* **4**(5): 667-677.

Ramos, J.M and del Amo, S. 1992 Enrichment planting in a tropical secondary forest in Veracruz, Mexico. *Forest Ecology and Management* **54**: 289-304.

Ramnarine, S. 1992 Effect of a slow release carbofuran on the Meliaceous shoot borer in Trinidad. Workshop paper of the 3rd Agricultural Research Seminar, NIHERST, Trinidad.

Rao, V. P. 1969 US PL-480 project: survey for natural enemies of *Hypsipyla robusta* Moore in India. Final technical report for the period July 25, 1964 to July 24, 1969. Commonwealth Institute of Biological Control, Trinidad.

Rao, V. P. and Bennett, F. D. 1969 Possibilities of biological control of the meliaceous shoot borers *Hypsipyla* spp. (Lepidoptera: Phycitidae). Technical Bulletin no. 12, pp. 61-81, Commonwealth Institute of Biological Control, Trinidad.

Revilla, A.V., Bonita, M.L. and Dimapilis, L.S. 1976a A yield prediction model for *Swietenia macrophylla* King plantations. *Pterocarpus* **2**(2): 172-179.

Revilla, A.V., Bonita, M.L. and Dimapilis, L.S. 1976b Tree volume functions for *Swietenia macrophylla* King plantations. *Pterocarpus* **2**(1): 1-7.

Rice, R.E., Gullison, R.E. and Reid, J.W. 1997 Can sustainable management save tropical forests? *Scientific American* (**April 1997**): 44-49.

Roberts, H. 1966 A survey of the important shoot, stem, wood, flower and fruit boring insects of the Meliaceae in Nigeria. Nigerian Forestry Information Bulletin no. 15 (New Series). 38 pp.

Roberts, H. 1968 An outline of the biology of *Hypsipyla robusta* Moore, the shoot borer of the Meliaceae (mahoganies) of Nigeria; together with brief comments on two stem borers and one other lepidopteran fruit borer also found in Nigerian Meliaceae. *Commonwealth Forestry Review* **47**(3): 225-232.

Roberts, H. 1973 Ambrosia beetles in Fiji. Report for the Forestry Department, Fiji.

Roberts, H. 1977 When ambrosia beetles attack mahogany trees in Fiji. *Unasylva* **29**(117): 25-28.

Rodan, B.D. and Campbell, F.T. 1996 CITES and the sustainable management of *Swietenia macrophylla* King. *Botanical Journal of the Linnean Society* **122**: 83-87.

Rodan, B.D., Newton, A.C. and Verissimo, A. 1992 Mahogany conservation: status and policy initiatives. *Environmental Conservation* 19(4): 331-342.

Rodriguez, O.P. 1996 Managing mahogany plantations. *Greenfields* 24(3): 8-15. Manila, Philippines.

Roldan, E.F. 1941 Nursery wilt of mahogany seedlings. *Philippine Journal of Forestry* 4(3): 267-278.

Rombouts, J. 1989 Mahogany and Jak plantation management plan. Unpublished report for the Forest Department, Sri Lanka.

Roovers, M. 1971 Observaciones sobre el ciclo de vida de *Hypsipyla grandella* (Zeller) en Barinitas, Venezuela. *Boletín del Instituto Forestal de Latino-Americano de Investigación y Capacitación* 38: 1-46.

Samarasinghe, S.J., Ashton, P.M.S., Gunatilleke, I.A.U.N. and Gunatilleke, C.V.S. 1995 Thinning guidelines for trees species of different successional status. *Journal of Tropical Forest Science* 8(1): 44-52.

San Buenaventura, P. 1958 Reforestation of *Imperata* waste lands in the Philippines. *The Philippine Journal of Forestry* 14(1-4): 67-76.

Sandiford, M. 1990 An account of the identification of existing *Swietenia macrophylla* populations in the Solomon Islands: a first step towards the improvement of *Swietenia macrophylla*. Forest Research Note 64-02/90, Forestry Division, Solomon Islands.

Sandom, J.H. 1978 The performance of exotic and indigenous species of minor importance in the Solomon Islands. Research Report S/1/78, Forestry Division, Solomon Islands.

Sandom, J.H. and Thayaparan, S. 1994 An Interim Management Plan for the Mahogany Forests of Sri Lanka. Unpublished report for the Forest Department, Colombo, Sri Lanka.

Sandom, J.H. and Thayaparan, S. 1995 A revision of the Interim Management Plan for the Mixed Mahogany Forests of Sri Lanka. Unpublished report for the Forest Department, Colombo, Sri Lanka.

Sankaran, K.V., Florence, E.J.J. and Sharma, J.K. 1984 Two new diseases of forest tree seedlings caused by *Sclerotium rolfsii* in India. *European Journal of Forest Pathology* 14(4/5): 318-320.

Santiago, B.R., Polito, J.C. and Cuevas, X.G. 1992 Dispersión de semillas y establecimiento de caoba (*Swietenia macrophylla*) despues de un tratamineto mecánico del sitio. *In*: L.K. Snook and A.B. de Jorgenson (eds.) Madera, Chicle, Caza y Milpa: Contribuciones al Manejo Integral de las Selvas de Quintana Roo, México. Memorias de Taller, 9 julio 1992, Chétumal, Quintana Roo, Mexico.

de Santis, L. 1972 Estudios sobre el barrenador *Hypsipyla grandella* (Zeller). IX. Un nuevo microgasterino néotropico (Hymenoptera, Braconidae) par sito de la larva. *Turrialba* 22(2): 222-224.

Schmidt, P.B. 1974 Sobre a profundidade ideal de semeadura do mogno, *Swietenia macrophylla* King. *Brasil Florestal* 5(17): 42-47.

Seroma, L. 1995 Information paper. Unpublished report for the Department of Forestry, Fiji.

Sheperd, K.R. 1969 A development plan for tree improvement in Fiji. *Cited in*: **Wright, H.L. 1969** Untitled. Unpublished report to the Department of Forestry, Suva, Fiji.

de Silva, D.W. 1952 Report of the Forest Department for the Year ended 31st December 1949. British Honduras.

Singh, S., and Bola, I. 1981 Diseases of plantation trees in Fiji islands. II. *Clitocybe tabescens*, root rot of mahogany, *Swietenia macrophylla* and *Pinus elliottii*. *Indian Journal of Forestry* **4**(2): 86-91.

Singh, S., Bola, I. and Kumar, J. 1980 Diseases of plantation trees in Fiji islands. I. Brown root rot of mahogany, *Swietenia macrophylla*. *Indian Forester* **106**(8): 526-532.

Smith, D.M. 1986 *The Practice of Silviculture (8th edition)*. John Wiley & Sons, Inc.

Snook, L.K. 1993 Stand dynamics of mahogany (*Swietenia macrophylla* King) and associated species after fire and hurricane in the tropical forests of the Yucatan Peninsula, Mexico. PhD thesis, School of Forestry and Environmental Studies, Yale University.

Snook, L.K., in press Sustaining harvests of mahogany (*Swietenia macrophylla* King) from Mexico's Yucatan Forests: past, present and future. *In*: Primack, Bray and Galletti (eds.) *Timber, Tourists and Temples: conservation and development in the Maya Forest of Belize, Guatemala and Mexico*. Island Press.

Soerianegara, I. and Lemmens, R.H.M.J. 1994 Timber trees: major commercial timbers. Plant Resources of South-East Asia 5(1), PROSEA, Bogor, Indonesia.

Somner, A. and Dow, T. 1978 Compilation of indicative growth and yield data of fast-growing exotic tree species planted in tropical and sub-tropical regions. FO:MISC/78/11, FAO, Rome, Italy.

Soubieux, J.M. 1983 Croissance et production du mahogany (*Swietenia macrophylla* King) en peuplements artificiels en Guadeloupe. École Nationale des Ingenieurs des Travaux des Eaux et Forêts, Direction Régionale pour la Guadeloupe, Office National des Forêts, Guadeloupe.

Spaan, W.J. 1909 Een menging van mahonie met soga. *Cited in*: **Indonesian Forestry Abstracts 1982** *Dutch literature until about 1960*. Abstract no. 1001. Wageningen, Netherlands.

Speight, M.R. and Wainhouse, D. 1988 *Ecology and Management of Forest Insects*. Oxford Science Publications, Clarendon Press, Oxford.

SRD. 1978 Silvicultural Research Division Annual Report. Department of Forestry, Suva, Fiji.

SRD. 1989 Silvicultural Research Division Annual Report. Department of Forestry, Suva, Fiji.

SRD. 1990 Silvicultural Research Division Annual Report. Department of Forestry, Suva, Fiji.

SRD. 1992 Silvicultural Research Division Annual Report. Department of Forestry, Suva, Fiji.

SRD. 1993 Silvicultural Research Division Annual Report. Department of Forestry, Suva, Fiji.

SRD. 1994 Silvicultural Research Division Annual Report. Department of Forestry, Suva, Fiji.

SRD. 1995 Silvicultural Research Division Annual Report. Department of Forestry, Suva, Fiji.

Stevenson, D. 1927a Forest research in British Honduras. *Bulletin of the Imperial Institute* (London): **25**(3): 313-320.

Stevenson, D. 1927b Report of the Forest Trust. Clarion Press, British Honduras.

Stevenson, N.S. 1927c Silvicultural treatment of mahogany forests in British Honduras. *Empire Forestry Journal* **6**(2): 219-227.

Stevenson, N.S. 1929 Report of the Forest Trust. Clarion Limited, Regent St., British Honduras.
Stevenson, N.S. 1930 Report of the Forest Trust. Clarion Limited, Regent St., British Honduras.
Stevenson, N.S. 1931 Annual Report of the Forest Trust. The Angelus Press, British Honduras.
Stevenson, N.S. 1935 Report of the Forest Trust for the Period 1st April 1933 to 31st December 1934. The Government Printer, British Honduras.
Stevenson, N.S. 1936 Report of the Forest Trust for the Year 1935. The Government Printer, British Honduras.
Stevenson, N.S. 1937 Report of the Forest Department for the Year 1936. The Government Printer, British Honduras.
Stevenson, N.S. 1939 Report of the Forest Department for the Year 1938. The Government Printer, British Honduras.
Stevenson, N.S. 1940 Report of the Forest Department for the Year 1939. The Government Printer, British Honduras.
Stevenson, N.S. 1941 Abridged Report of the Forest Department for the Year 1940. The Government Printer, British Honduras.
Stevenson, N.S. 1942 Abridged Report of the Forest Departmen for the Year 1941. The Government Printer, British Honduras.
Stevenson, N.S. 1943 Abridged Report of the Forest Department for the Year 1942. The Government Printer, British Honduras.
Stevenson, N.S. 1944a Abridged Report of the Forest Department for the Year 1943. The Government Printer, British Honduras.
Stevenson, N.S. 1944b A Forest Regeneration Scheme for British Honduras. Report for the Forest Department, British Honduras.
Stocker, C.L. 1924 Report of the Forest Trust, British Honduras. Waterlow and Sons Ltd., London Wall, London.
Streets, R.J. 1962 *Exotic trees in the British Commonwealth*. Clarendon Press, Oxford.
Styles, B. T. 1972 The flower biology of the Meliaceae and its bearing on tree breeding. *Silvae Genetica* **21**(5): 175-182.
Styles, B. T. and Khosla, P. K. 1976 Cytology and reproductive biology of Meliaceae. *In*: J. Burley and B.T. Styles (eds.) *Tropical trees: variation, breeding and conservation*, pp. 61-67. Linnean Society, London.
Suazo, J.T. 1990 Efecto del uso de herbicida Goal en la germinación de semillas de seis especies forestales. *El Tatascan* 7(1-2): 30-33. Siguatepeque, Honduras.
Sukanto, M. 1969 Climate of Indonesia. *In*: H. Arakawa (ed.) *World Survey of Climatology: Climate of Northern and Eastern Asia. Vol. 8*. Elsevier Publishing Co., Amsterdam.
Suratmo, F.G. 1979 Infestation of the leading shoots of mahogany (*Swietenia macrophylla* King) by *Hypsipyla robusta* Moore in West Java, Indonesia. Faculty of Forestry, Bogor Agriculture University, Indonesia.
Survey Department 1988 *The National Atlas of Sri Lanka*. Survey Department, Sri Lanka.
Taylor, G.F. 1977 The silviculture of *Swietenia* and *Cedrela* spp.: an overview of the problems and prospects of successful regeneration and development. Unpublished report for Dr C.DeZeeuw, College of Environmental Science and Forestry, Syracuse, New York.

TFF. 1994 Mahogany workshop: review and implications of the Convention of International Trade on Endangered Species. Tropical Forest Foundation, Virginia, USA.

Thorenaar, A. 1941 Over den cultuuraanleg in Bagelen. *Cited in*: **Indonesian Forestry Abstracts 1982** *Dutch literature until about 1960*. Abstract no. 870. Wageningen, Netherlands.

Tillier, S. 1995 Le mahogany grandes feuilles en Martinique. *Bois et Forêts des Tropiques* **244**(2): 55-66.

Tillmans, H.J. 1964 Apuntos bibliográphicos sobre *Hypsipyla grandella* Zeller. *Boletín del Instituto Forestal de Latino-Americano de Investigación y Capacitación* **16**: 82-92.

Tisseverasinghe, A.E.K. and Satchithananthan, S. 1957 The management of Sundapola plantation. *Ceylon Forester* **3**(1): 82-93.

Tito, B.V. and Rodriguez, M. 1988 (date uncertain) Estudio silvicultural preliminar de la caoba en la zona de Tingo María. Documentos de Trabajo no. 11, Avances de la Silvicultura en la Amazonia Peruana. República del Perú Instituto Nacional de Desarrollo, Perú.

Tompsett, P.B. 1994 Capture of genetic resources by collection and storage of seed: a physiological approach. *In*: R.R.B. Leakey and A.C. Newton (eds.) *Tropical trees: the potential for domestication and the rebuilding of forest resources*. Proceedings of the ITE Symposium no. 29 and ECTF Symposium no. 1, Heriot-Watt University, Edinburgh. HMSO, London.

Tropical Forest Experiment Station 1950 Tenth Annual Report for 1949. *Caribbean Forester* **11**(2): 59-80.

Tropical Forest Research Centre 1957 Seventeenth Annual Report. *Caribbean Forester* **18**(1): 1-11.

Tropical Forest Research Centre 1959 Nineteenth Annual Report. *Caribbean Forester* **20**(1): 1-10.

Troup, R.S. 1928 *Silvicultural systems*. Oxford Science Publications, Clarendon Press, Oxford.

Troup, R.S. 1932 *Exotic Forest Trees in the British Empire*. Oxford.

Valencia, L.C. and Ruiz, M.S.V. 1987 Alcances ecológico-silviculturales de la especie *Swietenia macrophylla* King. *Matero* **1**(1): 18-23.

Valera, M.Z. 1962 Study of the diameter growth of thinned large-leaf mahogany (*Swietenia macrophylla* King) stands in the Makiling National Park. *Philippine Journal of Forestry* **18**(1): 23-37.

Van Hall, C.J.J. 1918 Ziekten en plagen der cultuurgewassen in Nederlandsch-Indie in 1917. *Cited in*: **Indonesian Forestry Abstracts 1982** *Dutch literature until about 1960*. Abstract no. 1622. Wageningen, Netherlands.

Van Hall, C.J.J. 1919 Ziekten en plagen der cultuurgewassen in Nederlandsch-Indie in 1918. *Cited in*: **Indonesian Forestry Abstracts 1982** *Dutch literature until about 1960*. Abstract no. 1623. Wageningen, Netherlands.

Van Hall, C.J.J. 1922 Ziekten en plagen der cultuurgewassen in Nederlandsch-Indie in 1921. *Cited in*: **Indonesian Forestry Abstracts 1982** *Dutch literature until about 1960*. Abstract no. 1629. Wageningen, Netherlands.

Vanucci, C., Lange, C., Lhommet, G., Dupont, B., Davoust, D., Vauchot, B., Clement, J.L. and Brunck, F. 1992 An insect antifeedant limonoid from seed of *Khaya ivorensis*. *Phytochemistry* **31**: 3003-3004.

Varela, B. 1978 Documento mimeografiado presentado a la Escuala de Ingenieria Forestal del ITCR. *Cited in*: **Chavarria, E. and Valerio, R.V. 1993** Guía Preliminar de Parámetros Silviculturales para Apoyar los Proyectos de Reforestación en Costa Rica (Primera Versión). Ministerio de Recursos Naturales Energía y Minas, Dirección General Forestal, Costa Rica.

Vega, L. 1976 Influencia de la silvicultura en el comportamiento de *Cedrela* en Surinam. *In*: J.L. Whitmore (ed.) Studies on the shootborer *Hypsipyla grandella* (Zeller) Lep. Pyralidae. Volume III, pp. 26-49. IICA Miscellaneous Publications no. 101, CATIE, Turrialba, Costa Rica.

Vera, G. 1989 Estado de la investigación silvicultura de las plantaciones forestales en México. *In*: R. Salazar (ed.) Manejo y approvechamiento de plantaciones forestales con especies de uso múltiple. Actas Reunion IUFRO, April 1989, Guatemala.

Verissimo, A., Barreto P., Tarifa, R. and Uhl, C. 1995 Extraction of a high-value natural resource in Amazonia: the case of mahogany. *Forest Ecology and Management* **72**: 39-60.

Vincent, F. d'A. 1884 Mahogany in Ceylon. *Indian Forester* **10**: 156-157.

Vivekanandan, K. 1978 Retention of viability of mahogany seed through cold storage. *Sri Lanka Forester* **13**(3/4): 67-68.

Voogd, C.N.A. de 1948 De boscultuur van Janlappa. *Cited in*: **Indonesian Forestry Abstracts 1982** *Dutch literature until about 1960*. Abstract no. 976. Wageningen, Netherlands.

Wadsworth, F.H 1960 Datos de crecimiento de plantaciones forestales en México, Indias Occidentales, Centro y Sur America. *Cited in*: **Valencia, L.C. and Ruiz, M.S.V. 1987** Alcances ecológico-silviculturales de la especie *Swietenia macrophylla* King. *Matero* **1**(1): 18-23.

Wagner, M.R., Atuahene, S.K.N. and Cobbinah, J.R., 1991 *Forest Entomology in West tropical Africa: Forest Insects of Ghana*. Kluwer Academic Publishers, Dordrecht, Netherlands.

Wangaard, F.F. and Muschler, A.F. 1952 Properties and Uses of Tropical Woods III. Tropical Woods no. 98, pp. 148-149, School of Forestry, Yale University.

Waters, R.M. 1966 Annual Report of the Forest Department for the year 1965. Printing Department, Belize City, British Honduras.

Watkins, G. 1964 Mahogany: spacings and thinnings. Unpublished report no. F54/222 to the Department of Forestry, Suva, Fiji.

Watkins, H.B.R. and Gullison, R.E., in press Patterns of herbivory in a natural population of mahogany (*Swietenia macrophylla* King) saplings. *Biotropica*.

Weaver, P.L. 1987 Enrichment plantings in tropical America. *In*: J.C. Figueroa, F.H. Wadsworth and S. Branham (eds.) *Management of the Forests of Tropical America: prospects and technologies*, pp. 259-278. USDA Forest Service, Institute of Tropical Forestry, Puerto Rico.

Weaver, P.L. 1989 Taungya plantings in Puerto Rico: assessing the growth of mahogany and maria stands. *Journal of Forestry* **87**(3): 37-39.

Weaver, P.L. 1995 Forestry activities in the Caribbean basin with emphasis on Grenada. Unpublished report for the USDA Forest Service, pp. 259-278, Institute of Tropical Forestry, Puerto Rico.

Weaver, P.L. and Bauer, G.P. 1983 Crecimiento de caoba sembrada en líneas en la Sierra de Luquillo de Puerto Rico. Décimo Simposio de Recursos Naturales, 7 Diciembre 1983. Departamento de Recursos Naturales, San Juan, Puerto Rico.

Weaver, P.L. and Bauer, G.P. 1986 Growth, survival and shoot borer damage in mahogany plantings in the Luquillo Forest in Puerto Rico. *Turrialba* 36(4): 509-522.

Weaver, P.L. and Francis, J.K. 1988 Growth of teak, mahogany and Spanish cedar on St Croix, US Virgin Islands. *Turrialba* 38(4): 308-317.

Weaver, P.L. and Sabido, O.A. 1997 Mahogany in Belize: a historical perspective. General Technical Report IITF-2. USDA Forest Service, Institute of Tropical Forestry, Puerto Rico.

Webb, D.B., Wood, P.J., Smith, J.P. and Henmean, J.S. 1984 A guide to species selection for tropical and subtropical plantations. Tropical Forestry Publication 15, Commonwealth Forestry Institute, Oxford.

Weerawardane, N.D.R. 1996 Environmental effects on the growth of broadleaved trees introduced under pine stands in Sri Lanka. PhD thesis, Edinburgh University, UK.

Wescom, R.W. 1979 Silvicultural Research Division Annual Report 1978. Department of Forestry, Fiji.

Whitman, D. no date Provisional yield tables for Honduras mahogany, *Swietenia macrophylla*, in St. Lucia. Unpublished report for the Forestry Department, Castries, St. Lucia.

Whitmore, J.L. (ed.) 1976a *Studies on the shoot borer* Hypsipyla grandella *(Zeller) Lep. Pyralidae. Vol. II.* IICA Miscellaneous Publications No 101, CATIE, Turrialba, Costa Rica.

Whitmore, J.L. (ed.) 1976b *Studies on the shoor borer Hypsipyla grandella (Zeller) Lep. Pyralidae. Vol. III.* IICA Miscellaneous Publications No 101, CATIE, Turrialba, Costa Rica.

Whitmore, J.L. 1978 *Cedrela provenance trial in Puerto Rico and St. Croix: establishment phase.* USDA Forest Service Research Note ITF-16, Institute of Tropical Forestry, Puerto Rico.

Whitmore, J.L. 1983 *Swietenia macrophylla* and *S. humilis* (Caoba, mahogany). *In*: D.H. Janzen (ed.) Costa Rica Natural History, pp. 331-333. University of Chicago Press, Chicago and London.

Whitmore, J.L. 1992 An introduction to *Swietenia* with emphasis on silvics and silviculture. *In*: Mahogany Workshop: review and implications of CITES, 3-4 February. Tropical Forest Foundation, Washington D.C.

Whitmore, J.L. and Hinojosa, G. 1977 Mahogany (*Swietenia*) hybrids. Forest Service Research Paper ITF-23. Institute of Tropical Forestry, Río Piedras, Puerto Rico.

Whitmore, T.C. 1984 *Tropical Rain Forests of the Far East (2nd edition)*. Clarendon Press, Oxford.

Wilkins, R.M. 1972 Suppression of the shoot borer *Hypsipyla grandella* (Zeller) (Lepidoptera, Phyticidae) with controlled release insecticides. PhD thesis, University of Washington. 103 pp.

Wilkins, R.M., Allan, G.G. and Gara, R.I. 1976 Protection of Spanish cedar with controlled release insecticides. *In*: J.L. Whitmore (ed.) Studies on the shoot borer *Hypsipyla grandella* (Zeller) Lep. Pyralidae. Vol. III, pp. 63-70. IICA Miscellaneous Publications no. 101, CATIE, Turrialba, Costa Rica.

Wilson, P.H. 1986 A review of the occurrence of *Hypsipila robusta* in Solomon Island Forestry. Forestry Research Note 26/7/86, Forestry Division, Solomon Islands.

Winanto 1958 Over de aanplant van *Pinus merkusii* op Java. *Cited in*: **Indonesian Forestry Abstracts 1982** *Dutch literature until about 1960*. Abstract no. 654. Wageningen, Netherlands.

Wind, R. 1920 Eenige aanteekeningen omtrent een brand in culturen van verschillen. *Cited in*: **Indonesian Forestry Abstracts 1982** *Dutch literature until about 1960*. Abstract no. 476. Wageningen, Netherlands.

Wind Hzn, R. 1926 Bescherming van wild en bosch. *Cited in*: **Indonesian Forestry Abstracts 1982** *Dutch literature until about 1960*. Abstract no. 162. Wageningen, Netherlands.

Wolffsohn, A.L.A. 1961 An experiment concerning mahogany germination. *Empire Forestry Review* **40**(1): 71-72.

Wormald, T.J. 1992 Mixed and pure forest plantations in the tropics and subtropics. FAO Forestry Paper 103, Food and Agriculture Organisation, Rome, Italy.

Wright, H.L. 1968 Untitled. Unpublished report for the Department of Forestry, Fiji.

Yao, C.E. 1981 Survival and growth of mahogany seedlings under fertilized grassland conditions. *Sylvatrop* **6**(4): 203-217.

Yamazaki, S., Ikeda, T., Taketani, A., Pacheco, C.V. and Sato, T. 1992 Attack by the mahogany shoot borer, *Hypsipyla grandella* Zeller, on the Meliaceous trees in the Peruvian Amazon. *Applied Entomological Zoology* **27**(1): 31-38.

Yared, J.A.G. and Carpanezzi, A.A. 1981 Conversao de capoeira alta da Amazonia em povoamento de producao madeireira: o metoda 'recru' e especies promissoras. Boletím de Pesquisa no. 25, EMBRAPA, CPATU, Brazil.

Yaseen, M. and Bennett, F. D. 1975 Resumé of investigations on *Hypsipyla* and the release of parasites for their control in Trinidad during the period 1966-1975. Unpublished report for the Commonwealth Institute of Biological Control, Trinidad.

Zabala, N.Q. 1977 Field grafting of yemane, *Gmelina arborea* Roxb. and large leaf mahogany, *Swietenia macrophylla* King. *Pterocarpus* **3**(1): 81-86.

Zobel, B. and Talbert, J. 1984 Applied forest tree improvement. John Wiley & Sons, New York.

de Zoysa, N.D., Gunatilleke, C.V.S. and Gunatilleke, I.A.U.N. 1986 Vegetation studies of a skid-trail planted with mahogany in Sinharaja. *Sri Lanka Forester* **17**(3/4): 142-147.

Appendices

Appendix I Thinning regimes for mahogany stands

Table I-A. Thinning regime for mahogany stands in Martinique for an average site class (adapted from Tillier 1995)

Age (years)	Prescribed stocking (stems ha^{-1})	Predicted mean dbh (cm)	Predicted basal area before thinning (m^2 ha^{-1})	Predicted basal area after thinning (m^2 ha^{-1})
5	1600	nda	nda	nda
15	1000	nda	nda	nda
21	600	25.6	45.2	30.8
28	350	33.4	45.7	30.6
35	225	43.6	42.9	33.7
42	200	50.4	42.5	40.1
50	0	54.9	47.6	0

nda = no details available

Table I-B. Thinning regime for mahogany stands in Indonesia for an average site class (adapted from Anon. 1975)

Age (years)	Prescribed stocking (stems ha^{-1})	Predicted mean dbh (cm)	Predicted basal area before thinning (m^2 ha^{-1})	Predicted basal area after thinning (m^2 ha^{-1})
10	1070	12.4	-	12.9
20	500	27.2	62.2	29.1
30	325	36.5	52.3	34.0
40	240	43.5	48.3	35.7
50	181	49.8	46.7	35.3
60	140	55.5	43.8	33.9

Table I-C. Thinning regime for mahogany stands in Sri Lanka (adapted from Sandom and Thayaparan 1995)

Age (years)	Prescribed stocking (stems ha^{-1})	Predicted mean dbh (cm)	Predicted basal area before thinning (m^2 ha^{-1})	Predicted basal area after thinning (m^2 ha^{-1})
5	2500	5	-	4.9
15	900	12	28.3	10.2
25	600	20	28.3	18.8
35	270	30	42.4	19.1
45	100	40	33.9	12.6
55	30-35*	50	19.6	5.9-6.9
60	0	55	7.7	0

* After a regeneration felling to encourage natural regeneration

Appendix II Stocking of mahogany stands based on dbh-crown diameter relationships

Table II-A. Stocking of mahogany stands based on predicted crown diameters (adapted from Dawkins 1963)

Mean dbh (cm)	Predicted crown diam. (m)	Stocking for 78.5% cover (stems ha^{-1})	Basal area (m^2 ha^{-1})	Stocking for 100% cover (stems ha^{-1})	Basal area (m^2 ha^{-1})
5	1.35	5487	10.8	6986	13.7
10	2.30	1890	14.8	2407	18.9
15	3.25	947	16.7	1205	21.3
20	4.20	567	17.8	722	22.7
25	5.15	377	18.5	480	23.6
30	6.10	269	19.0	342	24.2
35	7.05	202	19.4	256	24.6
40	8.00	156	19.6	199	25.0
45	8.95	125	19.9	159	25.3
50	9.90	102	20.0	130	25.5
55	10.85	85	20.2	108	25.7
60	11.80	72	20.4	91	25.9

Crown diameter = 0.4+0.19(dbh)

Table II-B. Stocking of mahogany stands in Fiji based on predicted crown diameters (adapted from Wright 1968)

Mean dbh (cm)	Predicted crown diam. (m)	Stocking for 78.5% cover (stems ha^{-1})	Basal area (m^2 ha^{-1})	Stocking for 100% cover (stems ha^{-1})	Basal area (m^2 ha^{-1})
5	1.03	9363	18.4	12002	23.6
10	1.96	2610	20.5	3314	26.0
15	2.88	1204	21.3	1535	27.1
20	3.81	691	21.7	877	27.6
25	4.73	447	21.9	569	27.9
30	5.65	313	22.1	399	28.2
35	6.58	231	22.2	294	28.3
40	7.50	178	22.4	226	28.4
45	8.43	141	22.4	179	28.5
50	9.35	114	22.4	146	28.6

Crown diameter = 0.11+0.185(dbh)

Note: since the largest sampled tree was 40 cm dbh, predictions of crown diameter are not made for trees above 50 cm

Table II-C. Stocking of mahogany stands in Sri Lanka based on predicted crown diameters (adapted from Samarasinghe et al. 1995)

Mean dbh (cm)	Predicted crown diam. (m)	Stocking for 78.5% cover (stems ha^{-1})	Basal area (m^2 ha^{-1})	Stocking for 100% cover (stems ha^{-1})	Basal area (m^2 ha^{-1})
5	1.79	3121	6.1	3973	7.8
10	2.81	1266	9.9	1612	12.7
15	3.80	693	12.2	882	15.6
20	4.76	441	13.9	562	17.7
25	5.68	315	15.2	395	19.4
30	6.58	231	16.3	294	20.8
35	7.45	180	17.3	229	22.1
40	8.29	146	18.3	185	23.3
45	9.09	121	19.2	154	24.5
50	9.87	103	20.2	131	25.7
55	10.62	89	21.1	113	26.8
60	11.34	78	22.1	99	28.0

Crown diameter = $0.7466+0.2125(dbh)-0.0006(dbh^2)$

Appendix III Data from single inventories of mahogany stands

The data are set out in Table III-A.

Abbreviations used in Table III-A:
A_M	Age of mahogany (years)
N_M	Stocking of mahogany (stems ha^{-1})
N_T	Total stocking (all species including mahogany) (stems ha^{-1})
D_{mM}	Mean dbh of mahogany (cm)
D_{dM}	Dominant dbh of mahogany (cm)
H_{mM}	Mean height of mahogany (m)
H_{dM}	Dominant height of mahogany (m)
G_M	Basal area of mahogany (m^2 ha^{-1})
G_T	Total basal area (all species including mahogany) (m^2 ha^{-1})
V_M	Volume of mahogany (m^3 ha^{-1})
V_{MAIM}	Volume mean annual increment of mahogany (m^3 ha^{-1} yr^{-1})

Notes on Table III-A:
1. Figures in italics were not included in original literature, but have been calculated on the basis of available information.
2. Where D_{mM} and N_M were available, V_M was calculated using a volume equation from Fiji (Wescom 1979). Although designed for poor sites in Fiji, the equation is considered to be most typical of the range of volume equations in the literature (see Section 9.3). Single tree volumes were multiplied by N_M to give V_M and divided by A_M to give V_{MAIM}.
3. No V_M were calculated for stands or plots with a $D_{mM} < 10$ cm.

Table III-A. Data from single inventories of mahogany stands

Country	Source	Sample size	A_M	N_M	N_T	D_{mM}	D_{dM}	H_{mM}	H_{dM}	G_M	G_T	V_M	V_{MAIM}
Belize	Hummel (1925)	-	12.0	112	225	31.5	-	16.2	-	8.7	17.5	47.0	3.9
Belize	Lamb (1966)	-	11.0	692	-	14.0	-	8.5	-	10.6	-	57.1	5.2
Belize	Lamb (1966)	-	12.0	550	-	17.3	-	11.0	-	12.9	-	69.3	5.8
Belize	Lamb (1966)	-	13.0	1125	-	22.9	-	-	-	46.2	-	248.5	19.1
Belize	Lamb (1966)	-	17.0	775	-	30.5	-	-	-	56.5	-	304.3	17.9
Belize	Lamb (1966)	-	20.0	675	-	24.1	-	-	-	30.9	-	166.1	8.3
Belize	Lamb (1966)	-	30.0	500	-	35.6	-	-	-	49.7	-	267.3	8.9
Belize	Lamb (1966)	-	11.0	375	-	10.9	-	-	-	3.5	-	18.9	1.7
Brazil	Anon. (1988b)	small stand	5.0	-	-	4.0	-	9.8	-	-	-	-	0.2
Brazil	Golfari & Caser (1977)	small stand	6.3	-	-	13.0	-	2.7	-	-	-	-	-
Brazil	Pedroso & Pereira (1971)	94 trees	11.0	1600	1600	9.8	-	10.0	-	12.1	-	-	8.1
Fiji	SRD (1990)	1 plot	34.0	260	260	35.6	-	10.6	-	25.9	25.9	158.3	4.6
Fiji	SRD (1990)	1 plot	55.0	80	80	76.4	-	27.2	-	36.7	36.7	225.4	4.1
Fiji	SRD (1990)	1 plot	34.0	255	255	41.4	-	30.8	-	34.3	34.3	203.9	5.9
Fiji	MSD (1996)	many plots	32.5	149	-	-	-	29.3	-	-	-	204.4	7.5
Fiji	MSD (1996)	many plots	32.0	104	-	-	-	-	-	-	-	99.6	3.7
Fiji	MSD (1996)	many plots	31.0	117	-	-	-	-	-	-	-	118.5	4.5
Fiji	Wright (1968)	1 plot	11.0	1750	1750	14.0	15.8	16.5	-	27.4	27.4	144.4	13.1
Fiji	Wright (1968)	1 plot	11.0	1775	1775	10.9	26.4	13.4	-	17.0	17.0	89.5	8.1
Fiji	Wright (1968)	1 plot	12.0	888	888	18.8	16.0	21.3	-	25.1	25.1	132.6	11.0
Fiji	Wright (1968)	1 plot	12.0	1550	1550	11.9	20.6	13.7	-	17.9	17.9	93.4	7.8
Fiji	Wright (1968)	1 plot	13.0	3075	3075	11.4	18.3	18.6	-	33.0	33.0	169.8	13.1
Fiji	Wright (1968)	1 plot	13.0	900	900	11.9	18.5	19.8	-	10.2	10.2	54.2	4.2
Fiji	Wright (1968)	1 plot	13.0	575	575	13.5	27.2	14.3	-	8.4	8.4	44.0	3.4
Fiji	Wright (1968)	1 plot	13.0	650	650	12.2	-	17.1	-	11.6	11.6	40.8	3.1
Guadeloupe	Soubieux (1983)	1 plot	32.0	238	309	34.4	43.5	22.6	25.2	22.2	24.7	204.6	6.4
Guadeloupe	Soubieux (1983)	1 plot	27.0	400	405	33.2	47.1	20.6	24.1	34.7	34.9	305.4	11.3
Guadeloupe	Soubieux (1983)	1 plot	33.0	275	315	44.1	53.9	24.6	25.5	42.0	43.6	398.7	12.1
Guadeloupe	Soubieux (1983)	1 plot	21.0	720	730	26.6	43.6	18.0	21.2	39.9	39.9	310.4	14.8
Guadeloupe	Soubieux (1983)	1 plot	31.0	678	693	31.3	48.3	20.4	26.7	52.2	52.7	463.7	15.0
Guadeloupe	Soubieux (1983)	1 plot	26.0	955	980	25.2	42.5	15.6	19.9	47.6	48.0	339.6	13.1
Guadeloupe	Soubieux (1983)	1 plot	28.0	618	729	22.8	35.2	16.3	19.4	25.1	26.4	181.1	6.5
Guadeloupe	Soubieux (1983)	1 plot	28.0	1125	1216	17.0	33.6	15.0	20.7	25.6	26.7	174.4	6.2

Appendices

Country	Source	Plot											
Guadeloupe	Soubieux (1983)	1 plot	25.0	350	350	35.3	48.7	20.3	21.6	34.3	34.3	287.8	11.5
Guadeloupe	Soubieux (1983)	1 plot	22.0	832	955	21.7	35.1	16.7	19.4	30.7	31.7	220.4	10.0
Guadeloupe	Soubieux (1983)	1 plot	22.0	530	565	25.8	35.6	17.6	20.5	27.8	28.8	211.2	9.6
Guadeloupe	Soubieux (1983)	1 plot	19.0	1022	1073	16.6	29.5	13.4	18.9	22.1	23.4	137.4	7.2
Guadeloupe	Soubieux (1983)	1 plot	31.0	827	827	22.3	33.0	17.0	20.0	32.4	32.4	232.5	7.5
Guadeloupe	Soubieux (1983)	1 plot	31.0	477	502	25.0	35.9	18.4	22.5	23.5	24.1	187.0	6.0
Guadeloupe	Soubieux (1983)	1 plot	22.0	761	791	17.6	25.8	14.9	18.5	18.6	19.5	122.2	5.6
Guadeloupe	Soubieux (1983)	1 plot	22.0	631	652	22.7	31.2	18.1	20.5	25.6	26.1	195.9	8.9
Guadeloupe	Soubieux (1983)	1 plot	22.0	759	849	16.1	28.0	14.5	19.7	15.4	15.9	100.8	4.6
Guadeloupe	Soubieux (1983)	1 plot	22.0	734	823	15.2	24.1	13.7	18.0	13.4	14.3	81.3	3.7
Guadeloupe	Soubieux (1983)	1 plot	34.0	271	281	33.2	39.9	24.0	25.6	23.4	23.6	219.1	6.4
Guadeloupe	Soubieux (1983)	1 plot	25.0	1065	1200	16.4	25.9	13.6	17.5	22.4	23.6	136.8	5.5
Guadeloupe	Soubieux (1983)	1 plot	28.0	1293	1326	17.3	27.0	13.7	16.6	30.3	30.6	180.7	6.5
Guadeloupe	Soubieux (1983)	1 plot	27.0	1153	1248	17.4	26.8	13.3	16.4	27.4	28.1	158.9	5.9
Guadeloupe	Soubieux (1983)	1 plot	19.0	769	813	15.5	23.7	13.0	17.2	15.0	15.9	87.9	4.6
Guadeloupe	Soubieux (1983)	1 plot	27.0	353	374	37.3	51.9	25.1	30.2	38.6	42.4	390.8	14.5
Guadeloupe	Soubieux (1983)	1 plot	35.0	223	245	32.8	39.7	24.7	26.4	18.8	19.4	179.3	5.1
Guadeloupe	Soubieux (1983)	1 plot	35.0	447	493	27.7	35.7	20.5	23.0	27.0	28.0	225.5	6.4
Guadeloupe	Soubieux (1983)	1 plot	28.0	1125	1200	16.5	27.7	11.6	16.1	24.0	24.7	125.4	4.5
Guadeloupe	Soubieux (1983)	1 plot	28.0	1096	1107	15.6	28.4	12.1	16.8	20.9	21.0	118.7	4.2
Guadeloupe	Soubieux (1983)	1 plot	31.0	372	409	19.8	27.3	13.4	15.6	11.5	12.1	68.6	2.2
Guadeloupe	Soubieux (1983)	1 plot	31.0	268	281	30.0	35.9	18.9	21.9	19.0	19.4	153.9	5.0
Guadeloupe	Soubieux (1983)	1 plot	21.0	826	1005	13.9	20.5	11.1	13.8	12.5	13.3	60.5	2.9
Guadeloupe	Soubieux (1983)	1 plot	18.0	734	779	21.3	30.4	18.5	19.9	26.2	26.6	198.0	11.0
Guadeloupe	Soubieux (1983)	1 plot	31.0	895	955	20.3	29.3	17.9	21.4	29.0	30.0	215.1	6.9
Guadeloupe	Soubieux (1983)	1 plot	34.0	540	540	27.6	36.0	21.7	25.2	32.3	32.3	283.2	8.3
Guadeloupe	Soubieux (1983)	1 plot	27.0	980	1055	14.2	21.5	10.1	13.3	15.4	16.4	71.0	2.6
Guadeloupe	Soubieux (1983)	1 plot	25.0	1095	1740	12.6	18.1	9.2	10.1	13.6	17.1	52.3	2.1
Guadeloupe	Soubieux (1983)	1 plot	34.0	585	690	24.7	34.3	17.6	18.9	28.0	29.8	205.6	6.0
Honduras	Hazlett & Montesinos (1980)	a few small plots	35.0	-	-	37.7	-	22.0	-	-	-	-	-
Honduras	James (1990)	10 trees	11.0	-	816	24.6	39.6	-	-	-	38.8	-	-
Honduras	James (1990)	8 trees	41.0	-	239	77.7	93.0	-	-	-	113.3	-	-
Honduras	James (1990)	7 trees	42.0	-	816	41.5	49.0	-	-	-	110.4	-	-
Honduras	Lamb (1966)	-	13.0	177.5	-	21.1	-	13.7	-	6.2	-	33.3	2.6
Honduras	Puerto (1983)	1 plot (59 trees)	35.0	295	295	32.8	45.5	20.9	25.4	24.9	24.9	134.1	3.8
Honduras	Puerto (1983)	1 plot (59 trees)	35.0	230	230	26.5	32.0	24.6	20.9	12.7	12.7	68.3	2.0
Indonesia	Lamb (1966)	-	10.0	-	-	11.4	-	15.9	-	-	-	-	-
Indonesia	Lamb (1966)	-	10.0	-	-	16.8	-	21.4	-	-	-	-	-

Table III-A (continued). Data from single inventories of mahogany stands

Country	Source	Sample size	A_M	N_M	N_T	D_{mM}	D_{dM}	H_{mM}	H_{dM}	G_M	G_T	V_M	V_{MAIM}
Indonesia	Lamb (1966)	-	11.0	-	-	13.7	-	21.4	-	-	-	-	-
Martinique	Lamb (1966)	-	16.0	-	-	30.5	-	24.4	-	-	-	-	-
Martinique	Lamb (1966)	-	17.0	-	-	17.8	-	26.8	-	-	-	-	-
Martinique	Lamb (1966)	-	36.0	-	-	53.3	-	30.5	-	-	-	-	-
Martinique	Marie (1949)	district average	1.0	-	-	2.9	-	1.9	-	-	-	-	-
Martinique	Marie (1949)	district average	2.0	-	-	4.1	-	3.3	-	-	-	-	-
Martinique	Marie (1949)	district average	4.0	-	-	8.6	-	5.3	-	-	-	-	-
Martinique	Marie (1949)	district average	6.0	-	-	15.3	-	7.1	-	-	-	-	-
Martinique	Marie (1949)	district average	10.0	-	-	23.6	-	12.8	-	-	-	-	-
Martinique	Marie (1949)	district average	15.0	-	-	33.1	-	18.5	-	-	-	-	-
Martinique	Marie (1949)	district average	20.0	-	-	46.5	-	20.3	-	-	-	-	-
Martinique	Marie (1949)	district average	25.0	-	-	62.1	-	22.5	-	-	-	-	-
Martinique	Lamb (1966)	-	32.0	-	-	50.8	-	30.5	-	-	-	-	-
Mexico	Lamb (1966)	-	7.0	825	-	5.8	-	5.5	-	2.2	-	11.9	1.7
Mexico	Lamb (1966)	-	8.0	5000	-	8.1	-	7.9	-	25.9	-	139.6	17.5
Mexico	Lamb (1966)	-	12.0	6075	-	8.9	-	5.5	-	37.7	-	202.9	16.9
Nicaragua	Lamb (1966)	-	7.0	600	-	10.9	-	9.2	-	5.6	-	30.2	4.3
Nicaragua	Lamb (1966)	-	9.0	2250	-	13.0	-	11.0	-	29.7	-	159.6	17.7
Nicaragua	Lamb (1966)	-	9.0	600	-	10.2	-	11.0	-	4.9	-	26.2	2.9
Nicaragua	Lamb (1966)	-	10.0	180	-	21.6	-	13.1	-	6.6	-	35.4	3.5
Nicaragua	Lamb (1966)	-	13.0	180	-	25.4	-	18.3	-	9.1	-	49.1	3.8
Nicaragua	Lamb (1966)	-	13.0	300	-	20.3	-	15.3	-	9.7	-	52.3	4.0
Nigeria	DFR (1963)	1 plot (69 trees)	2.3	-	-	-	-	2.1	-	-	-	-	-
Nigeria	DFR (1963)	1 plot (64 trees)	1.5	-	-	-	-	1.6	-	-	-	-	-
Nigeria	DFR (1963)	1 plot (64 trees)	1.5	-	-	-	-	2.3	-	-	-	-	-
Nigeria	DFR (1963)	1 plot (128 trees)	1.5	-	-	-	-	1.7	-	-	-	-	-
Panama	Lamb (1966)	-	11.0	225	-	21.8	-	13.7	-	8.4	-	45.4	4.1
Peru	Lamb (1966)	-	20.0	475	-	38.1	-	22.9	-	54.2	-	291.5	14.6
Philippines	Chinte (1952)	25 trees	36.0	-	156	50.3	-	21.3	-	-	-	123.2	3.4
Philippines	Chinte (1952)	25 trees	35.0	-	123	51.3	-	21.0	-	-	-	83.6	2.4
Philippines	Chinte (1952)	12 trees	22.0	-	228	39.9	-	19.0	-	-	-	84.4	3.8
Philippines	Chinte (1952)	32 trees	22.0	-	228	37.1	-	16.6	-	-	-	63.8	2.9
Philippines	Chinte (1952)	64 trees	17.0	-	400	25.9	-	19.1	-	-	-	80.0	4.7

Country	Reference	Sample									
Philippines	Chinte (1952)	37 trees	21.0	-	-	-	-	-	-	-	-
Philippines	Chinte (1952)	36 trees	21.0	-	-	-	-	-	-	-	-
Philippines	Chinte (1952)	38 trees	21.0	-	-	-	-	-	-	-	-
Philippines	Chinte (1952)	87 trees	18.0	-	-	-	-	-	-	-	-
Puerto Rico	Lamb (1966)	-	14.0	1200	-	35.3	14.3	-	24.1	-	129.9	9.3
Puerto Rico	Lamb (1966)	-	20.0	300	-	37.9	15.3	-	29.8	-	160.3	8.0
Puerto Rico	Lamb (1966)	-	24.0	-	-	23.3	14.7	-	-	-	-	-
Puerto Rico	Lamb (1966)	-	10.0	1850	-	20.9	13.6	-	-	-	-	-
Puerto Rico	Lamb (1966)	-	16.0	935	-	16.0	22.9	-	23.4	-	126.1	12.6
Puerto Rico	Lamb (1966)	-	17.0	1000	-	35.6	24.4	-	38.4	-	206.5	12.9
Puerto Rico	Lamb (1966)	-	26.0	562	-	48.3	30.5	-	32.4	-	174.5	10.3
Puerto Rico	Lamb (1966)	-	13.0	1850	-	12.7	-	-	48.2	-	259.2	10.0
Puerto Rico	Marrero (1950)	1 stand	6.0	-	-	22.9	-	-	21.6	-	116.4	9.0
Puerto Rico	Marrero (1950)	1 stand	10.0	-	-	20.3	-	-	-	-	-	-
Puerto Rico	TFRC (1957)	1 plot	13.0	280	-	33.0	-	-	-	25.5	35.4	2.7
Puerto Rico	TFRC (1957)	1 plot	13.0	180	-	12.2	-	-	-	19.5	26.0	2.0
Puerto Rico	TFRC (1957)	1 plot	13.0	190	-	7.6	-	-	-	19.5	20.5	1.6
Puerto Rico	TFRC (1957)	1 plot	13.0	190	-	10.4	-	-	-	18.6	18.0	1.4
Puerto Rico	TFRC (1958)	1 plantation	27.0	-	358	17.3	-	-	-	35.6	-	-
Puerto Rico	Weaver & Bauer (1986)	4 sites	9.0	-	-	18.5	-	-	-	-	70.0	7.8
Puerto Rico	Weaver & Bauer (1986)	10 sites	15.0	-	-	16.0	-	-	-	-	125.0	8.3
Puerto Rico	Weaver & Bauer (1986)	3 sites	26.0	-	-	15.0	-	-	-	-	234.0	9.0
Puerto Rico	Weaver & Bauer (1986)	1 stand (50%)	18.0	-	-	35.6	18.0	-	-	-	-	-
Puerto Rico	Weaver & Bauer (1986)	1 stand (50%)	14.0	-	-	25.2	9.8	-	-	-	-	-
Puerto Rico	Weaver & Bauer (1986)	1 stand (50%)	7.0	-	-	10.8	4.9	-	-	-	-	-
Sabah	Corpuz & Cockburn (1972)	1 plot	4.0	-	1329	4.3	-	-	-	-	-	-
Sabah	Corpuz & Cockburn (1972)	1 plot	3.0	-	1329	9.3	2.9	-	-	-	-	-
Sabah	Corpuz & Cockburn (1972)	1 plot	3.0	-	420	-	-	-	-	-	-	-
Sabah	Corpuz & Cockburn (1972)	1 plot	3.0	-	1076	4.9	-	-	-	-	-	-
Sabah	Corpuz & Cockburn (1972)	1 plot	3.0	-	1329	4.6	-	-	-	-	-	-
Sabah	Corpuz & Cockburn (1972)	1 plot	3.0	-	1682	7.8	-	-	-	-	-	-
Sabah	Corpuz & Cockburn (1972)	1 plot	3.0	-	997	6.4	-	-	-	-	-	-
Sabah	Corpuz & Cockburn (1972)	1 plot	3.0	-	478	5.4	-	-	-	-	-	-
Sabah	Forest Department (no date.)	1 plot	11.8	-	-	14.7	-	-	-	-	-	-
Solomon Is.	Chaplin (1993)	1 plot	2.0	-	-	-	-	5.5	-	-	-	-
Solomon Is.	Chaplin (1993)	1 plot	3.0	-	-	-	-	8.2	-	-	-	-
Solomon Is.	Chaplin (1993)	1 plot	3.0	-	-	-	-	8.0	-	-	-	-
Solomon Is.	Chaplin (1993)	1 plot	3.0	-	-	-	-	7.5	-	-	-	-
Solomon Is.	Chaplin (1993)	1 plot	4.0	-	-	-	-	10.8	-	-	-	-

Table III-A (continued). Data from single inventories of mahogany stands

Country	Source	Sample size	A_M	N_M	N_T	D_{mM}	D_{dM}	H_{mM}	H_{dM}	G_M	G_T	V_M	V_{MAIM}
Solomon Is.	Chaplin (1993)	1 plot	5.0	-	-	-	-	-	11.1	-	-	-	-
Solomon Is.	Chaplin (1993)	1 plot	17.0	-	-	22.0	-	-	17.0	-	-	-	-
Solomon Is.	Chaplin (1993)	1 plot	30.0	-	-	58.0	-	-	29.5	-	-	-	-
Solomon Is.	Chaplin (1993)	1 plot	31.0	-	-	55.0	-	-	29.0	-	-	-	-
Solomon Is.	Sandom (1978)	1 plot	19.0	-	-	46.3	-	28.0	-	-	-	-	-
Solomon Is.	Sandom (1978)	1 plot	19.0	-	-	26.5	-	25.3	-	-	-	-	-
Sri Lanka	Anon. (1935)	1 plot (~50 trees)	10.0	-	-	17.0	-	10.4	-	-	-	-	-
Sri Lanka	Anon. (1935)	1 plot (~50 trees)	20.0	-	-	35.5	-	18.6	-	-	-	-	-
Sri Lanka	Anon. (1935)	1 plot (~50 trees)	30.0	-	-	53.4	-	25.0	-	-	-	-	-
Sri Lanka	Anon. (1935)	1 plot (~50 trees)	40.0	-	-	68.7	-	30.2	-	-	-	-	-
Sri Lanka	Perera (1955)	a few plots	15.5	-	-	-	-	16.4	-	-	-	-	-
St. Croix	Lamb (1966)	-	52.0	-	-	36.1	-	15.2	-	-	-	-	-
Trinidad	Lamb (1966)	-	10.0	-	-	22.6	-	18.0	-	-	-	-	-
Trinidad	Marshall (1939)	0.4 ha	16.0	-	290	22.6	-	18.0	-	-	11.6	-	-
Vanuatu	Leslie (1994)	1 plot	5.0	-	-	6.0	-	-	19.5	-	-	-	-
Vanuatu	Leslie (1994)	1 plot	5.0	-	-	13.0	-	-	5.5	-	-	-	-
Vanuatu	Leslie (1994)	1 plot	6.0	-	-	9.0	-	-	11.5	-	-	-	-
Vanuatu	Leslie (1994)	1 plot	6.0	-	-	16.0	-	-	7.5	-	-	-	-
Vanuatu	Leslie (1994)	1 plot	8.0	-	-	-	-	-	16.0	-	-	-	-
Vanuatu	Leslie (1994)	1 plot	10.0	-	-	18.0	-	-	10.0	-	-	-	-
Vanuatu	Leslie (1994)	1 plot	10.0	-	-	18.0	-	-	14.5	-	-	-	-
Vanuatu	Leslie (1994)	1 plot	10.0	-	-	22.0	-	-	16.0	-	-	-	-
Vanuatu	Leslie (1994)	1 plot	13.0	-	-	38.0	-	-	15.0	-	-	-	-
Vanuatu	Leslie (1994)	1 plot	20.0	-	-	-	-	-	20.0	-	-	-	-
Vanuatu	Leslie (1994)	1 plot	20.0	-	-	36.0	-	-	25.5	-	-	-	-
Vanuatu	Leslie (1994)	1 plot	21.0	-	-	-	-	-	24.5	-	-	-	-
Vanuatu	Leslie (1994)	1 plot	21.5	-	-	53.0	-	-	26.0	-	-	-	-
Vanuatu	Leslie (1994)	1 plot		-	-		-	-	27.0	-	-	-	-

TFRC = Tropical Forest Research Centre

Appendix IV Data from continuous inventories of mahogany stands

The data are set out in Table IV-A.

Abbreviations used in Table IV-A:
P_n Local plot code number
P_a Plot area (ha)
R_n Number of plot remeasurements
A_M Age range of mahogany (years)
N_M Stocking range of mahogany (stems ha^{-1})
D_{mM} Mean dbh range of mahogany (cm)
D_{dM} Dominant dbh range of mahogany (cm)
H_{mM} Mean height range of mahogany (m)
H_{dM} Dominant height range of mahogany (m)
G_{fM} Basal area of mahogany at the time of the final measurement (m^2 ha^{-1})
G_{fT} Total basal area (all species including mahogany) at the time of final measurement (m^2 ha^{-1})
V_{fM} Volume of mahogany at the time of the final measurement (m^3 ha^{-1})
V_{MAI} Volume mean annual increment of mahogany (m^3 ha^{-1} yr^{-1})

Notes on Table IV-A:
Certain data from the following countries were not considered suitable for further analysis:
1. Predominant (pertaining to the 20 trees of largest dbh) parameters from Fiji are incompatible with dominant (pertaining to the 100 trees of largest dbh) parameters. Most forest managers regard dominant parameters as standard. Comparison of predominant parameters from Fiji with dominant parameters from other countries would not have been an informative excercise.
2. Data from the Mexican plot were associated with negative increment due to a heavy thinning in 1986 (Parraguirre pers. comm.).
3. Analysis of stand parameters from St. Lucia and associated literature (Whitman no date) revealed that recent crown thinnings had produced considerable negative dbh and height increment in some plots. It appears that the inclusion of in-growth in later inventories has reduced mean stand parameters.
4. The permanent sample plots in Sri Lanka were located in stands containing a mixture of mahogany and *Artocarpus integrifolia* (Forest Department archives). The effect of crown thinning of mahogany to favour *A. integrifolia* may have had a considerable impact on the overall productivity of mahogany.

Table IV-A. Data from continuous inventories of mahogany stands

Country	Location	P_n	P_a	R_n	A_M	N_M	D_{mM}	D_{dM}	H_{mM}	H_{dM}	G_{fM}	G_{TT}	V_{fM}	V_{MAI}	Source
Belize	Columbia River	CR15	0.2	11	7-38[a]	321-252	5.7-29.9	-	-	8.5-27.7	20.2	-	189	5.0	Ennion (1996)
Belize	Columbia River	CR19	0.2	11	7-36[a]	119-119[b]	4.1-29.4	-	-	5.0-25.3	10.4	-	94	2.6	Ennion (1996)
Belize	Columbia River	CR20	0.2	10	5-36[a]	119-138[b]	4.9-26.2	-	-	6.1-24.4	8.7	-	75	2.1	Ennion (1996)
Belize	Columbia River	CR26A	0.12	7	12-38[a]	305-288	10.1-30.1	-	-	17.6-29.3	23.2	-	237	6.2	Ennion (1996)
Belize	Silk Grass	I.G-1	-	3	2-9	-	-	-	1.1-3.9	-	-	-	-	-	Stevenson (1936)
Belize	Silk Grass	I.G-2	-	3	1-8	-	-	-	0.7-6.4	-	-	-	-	-	Stevenson (1936)
Belize	Silk Grass	I.G-3	-	2	3-7	-	-	-	3.5-7.0	-	-	-	-	-	Stevenson (1936)
Belize	Silk Grass	I.G-4	-	2	3-7	-	-	-	2.4-5.3	-	-	-	-	-	Stevenson (1936)
Belize	Silk Grass	I.10.1	-	3	1-8	-	-	-	0.2-4.0	-	-	-	-	-	Stevenson (1936)
Belize	Columbia River	-	-	2	5-7.5	-	-	-	5.0-6.5	-	-	-	-	-	Stevenson (1936)
Belize	Columbia River	-	-	2	5-7.5	-	-	-	4.3-5.2	-	-	-	-	-	Stevenson (1936)
Belize	Columbia River	-	-	2	4-6.5	-	-	-	3.8-8.3	-	-	-	-	-	Stevenson (1936)
Belize	Columbia River	-	-	2	4-6.5	-	-	-	3.4-3.3	-	-	-	-	-	Stevenson (1936)
Belize	Columbia River	-	-	2	4-6.5	-	-	-	3.4-5.3	-	-	-	-	-	Stevenson (1936)
Belize	Columbia River	-	-	2	4-6.5	-	-	-	3.7-4.9	-	-	-	-	-	Stevenson (1936)
Belize	Columbia River	-	-	2	3-5.5	-	-	-	2.4-5.8	-	-	-	-	-	Stevenson (1936)
Belize	Columbia River	-	-	2	3-5.5	-	-	-	2.8-5.5	-	-	-	-	-	Stevenson (1936)
Cuba	-	-	0.07	7	1-7	-	12.6-19.7	-	1-6	-	-	-	-	-	Aguila et al. (1979)
Dominica	Cabrits	-	-	6	8-15	-	-	-	-	-	-	-	-	-	Anon. (1979)
Dominica	Cabrits	-	-	6	8-15	-	10.7-15.3	-	-	-	-	-	-	-	Anon. (1979)
Fiji	Nukurua	76	0.2	2	15-26	163-163	35.4-49.7	45.6-54.2[d]	22.5-29.5	25.3-33.0[d]	31.6	31.6	192	7.4	SRD (1978)[f]
Fiji	Nukurua	77	0.2	2	14-25	119-109	32.2-48.3	41.8-48.2[d]	21.8-30.0	24.0-35.6[d]	19.9	22.6	122	4.9	SRD (1978)[f]
Fiji	Nukurua	78	0.2	2	13-24	173-158	28.8-48.5	39.5-54.8[d]	21.9-30.6	23.0-32.6[d]	29.2	29.4	179	7.5	SRD (1978)[f]
Fiji	Nukurua	79	0.2	2	13-24	119-99	23.3-47.2	35.4-47.3[d]	18.1-30.2	20.4-32.7[d]	17.5	20.2	106	4.4	SRD (1978)[f]
Fiji	Nukurua	80	0.2	2	10-21	168-168	25.2-44.1	31.1-49.6[d]	18.1-27.9	20.7-28.8[d]	25.8	25.9	156	7.4	SRD (1978)[f]
Fiji	Nukurua	81	0.2	2	10-21	217-193	25.8-42.3	32.7-51.2[d]	19.8-30.0	21.7-30.5[d]	27.1	29.9	164	7.8	SRD (1978)[f]
Fiji	Nukurua	82	0.2	10	5-21[a]	217-202	19.0-38.0	27.7-46.1[d]	15.2-24.3	11.0-27.0[d]	22.9	24.2	139	6.6	SRD (1978)[e f]
Fiji	Nukurua	83	0.2	10	5-21[a]	203-188	19.8-41.9	27.5-49.4[d]	15.2-28.0	9.5-27.0[d]	25.9	26.3	153	7.3	SRD (1978)[e f]
Fiji	Nukurua	84	0.2	5	5-26[a]	153-149	9.4-49.9	42.0-56.5[d]	25.0-32.0	13.1-33.5[d]	29.2	30.9	191	7.3	SRD (1978)[f g]
Fiji	Nukurua	85	0.2	11	6.5-26[a]	183-183	7.6-35.6	32.6-41.6[d]	19.3-27.4	9.8-28.2[d]	18.2	18.2	110	4.2	SRD (1978)[e f g]
Fiji	Nukurua	86	0.2	2	13-24	247-213	26.7-43.3	43.0-54.9[d]	21.0-31.2	23.8-33.5[d]	31.3	34.3	193	8.0	SRD (1978)[f]

Country	Location	ID											Reference
Fiji	Nukurua	87	0.2	2	12-23	751-713	22.2-30.6	39.7-50.0[d]	19.9-29.9	24.5-30.3[d]	52.4	60.3	314	13.7	SRD (1978)[f]
Fiji	Nukurua	88	0.2	2	12-23	361-287	25.2-39.6	38.7-51.9[d]	20.5-31.2	24.8-31.5[d]	35.3	49.7	214	9.3	SRD (1978)[f]
Fiji	Nukurua	89	0.2	5	5.3-25[a]	158-153	7.4-44.7	40.8-49.8[d]	21.0-29.5	10.1-30.7[d]	24.1	26.2	146	5.8	SRD (1978)[g]
Fiji	Nukurua	90	0.2	3	4.3-6.4	335-330	7.6-11.7	9.4-14.7[d]	-	10.1-13.7[d]	3.6	-	-	-	Wright (1968)
Fiji	Nukurua	91	0.2	5	5.3-26[a]	217-203	7.1-42.0	36.8-49.7[d]	21.2-28.1	11.0-28.2[d]	28.1	31.3	170	6.5	SRD (1978)[g]
Fiji	Nukurua	92	0.2	4	4.3-25[a]	217-208	4.3-38.3	32.6-46.7[d]	19.0-29.2	11.3-24.5[d]	24.0	27.1	152	6.1	SRD (1978)[g]
Fiji	Nukurua	93	0.2	3	4.3-6.4[a]	370-340	6.4-10.7	10.9-12.7[d]	7.9-10.1	11.6-13.7[d]	3.0	-	-	-	Wright (1968)
Fiji	Nukurua	94	0.2	3	4.4-6.4[a]	365-335	6.4-10.4	10.2-12.4[d]	7.0-8.5	8.2-13.7[d]	2.9	-	-	-	Wright (1968)
Fiji	Nukurua	95	0.2	3	4.4-6.4[a]	365-365	6.9-10.9	11.7-13.7[d]	-	-	3.4	-	-	-	Wright (1968)
Fiji	Nukurua	19	0.4	5	24-28	180-178	26.2-30.7	-	-	-	13.2	-	-	-	Wright (1968)
Fiji	Nukurua	20	0.4	6	24-29	180-168	26.2-32.0	-	-	-	13.5	-	-	-	Wright (1968)
Fiji	Nukurua	28	0.4	8	8-17[a]	232-185	15.0-36.8	38.1-42.9[d]	-	-	19.7	-	-	-	Wright (1968)
Fiji	Nukurua	30	0.4	5	27-32[a]	242-198	37.1-43.4	47.0-49.5[d]	-	-	29.3	-	-	-	Wright (1968)
Fiji	Nukurua	33	0.4	2	14-16.5	255-255	23.6-26.9	26.4-30.0[d]	-	-	14.4	-	-	-	Wright (1968)
Mexico	Quintana Roo	-	1.5	5	10-28[a]	705-572	6.3-20.9	-	5.1-12.7	-	-	-	-	-	Cuevas et al.[h]
Philippines	Ilocos Sur	-	-	2	2-7	-	-	-	1.0-4.0	-	-	-	-	-	Nastor (1957)
Philippines	Luzon	A	-	6	8-14	-	9.8-16.4	-	-	-	-	-	-	-	Ponce (1933)
Philippines	Luzon	1-A	-	6	8-14	-	14.1-29.3	-	-	-	-	-	-	-	Ponce (1933)
Philippines	Luzon	3-A	-	6	7-13	-	9.9-19.3	-	-	-	-	-	-	-	Ponce (1933)
Philippines	Luzon	C	-	6	9-15	-	14.4-22.0	-	-	-	-	-	-	-	Ponce (1933)
St. Lucia	Island-wide	1	0.05	3	20	1420-940	17.4-24.3	-	12.4-22.2	15.7-24.3	48.3	48.3	-	-	Andrew (1994)
St. Lucia	Island-wide	2	0.05	3	19	1550-840	16.4-20.9	-	14.5-18.0	18.1-29.2	37.3	37.3	-	-	Andrew (1994)
St. Lucia	Island-wide	3	0.05	3	19	2020-1340	13.3-16.6	-	12.9-18.1	15.5-22.3	31.7	31.7	-	-	Andrew (1994)
St. Lucia	Island-wide	16	0.05	3	23	440-320	32.2-41.8[c]	-	24.0-21.9[c]	23.1-30.3	24.6	24.6	-	-	Andrew (1994)
St. Lucia	Island-wide	17	0.05	3	14	680-480	28.6-37.0	-	26.6-30.3	27.0-31.5	55.5	55.5	-	-	Andrew (1994)
St. Lucia	Island-wide	18	0.05	3	19	700-420	23.8-30.3	-	17.4-22.0	19.3-22.7	34.6	34.6	-	-	Andrew (1994)
St. Lucia	Island-wide	19	0.05	3	18	560-380	23.9-29.3[c]	-	15.4-21.7	18.2-22.8	27.0	27.0	-	-	Andrew (1994)
St. Lucia	Island-wide	22	0.05	3	20	580-380	29.9-37.9	-	20.0-26.5	14.2-30.5	46.7	46.7	-	-	Andrew (1994)
St. Lucia	Island-wide	23	0.05	3	18	1180-860	15.8-25.9	-	12.3-19.2	13.4-21.4	48.6	48.6	-	-	Andrew (1994)
St. Lucia	Island-wide	25	0.05	3	12	600-480	23.0-29.3	-	13.5-21.2	16.1-20.4	34.3	34.3	-	-	Andrew (1994)
St. Lucia	Island-wide	26	0.05	3	15	720-560	22.2-24.7[c]	-	15.5-20.9	14.6-21.7	28.4	28.4	-	-	Andrew (1994)
St. Lucia	Island-wide	29	0.05	3	16	760-540	23.8-26.7[c]	-	16.5-22.3	28.4-23.7[c]	31.6	31.6	-	-	Andrew (1994)
St. Lucia	Island-wide	31	0.05	3	10	560-640	12.9-15.5	-	8.1-13.0	9.2-14.3	13.8	13.8	-	-	Andrew (1994)
Solomon Is.	Gizo	-	-	4	1.7-9.2[a]	-	5.6-10.7	-	3.3-10.6	-	-	-	-	-	Sandom (1978)
Solomon Is.	Ringi	-	-	3	2.3-4.5	-	5.8-8.4	-	5.3-8.3	-	-	-	-	-	Sandom (1978)

Table IV-A (continued). Data from continuous inventories of mahogany stands

Country	Location	P_n	P_a	R_n	A_M	N_M	D_{mM}	D_{dM}	H_{mM}	H_{dM}	G_{fM}	G_{rT}	V_{fM}	V_{MAI}	Source
Solomon Is.	Mount Austen	-	-	3	12-18.5	-	35.6-47.0	-	18.9-27.9	-	-	-	-	-	Sandom (1978)
Solomon Is.	Allardyce	-	-	3	1-4.3[a]	-	8.2-10.4	-	2.5-11.4	-	-	-	-	-	Sandom (1978)
Solomon Is.	Viru	-	-	4	1-4[a]	-	4.7-8.8	-	1.7-9.5	-	-	-	-	-	Sandom (1978)
Sri Lanka	Dawatagolla	3A	0.4	13	13-25	895-247	14.7-24.6	20.6-35.6	-	-	11.7	-	-	-	FD archives
Sri Lanka	Dawatagolla	3B	0.4	12	13-24	652-168	14.8-24.4	19.7-32.5	-	-	7.9	-	-	-	FD archives
Sri Lanka	Dawatagolla	Q16	0.2	8	19-26	360-197	19.7-24.2	27.0-31.9	-	-	9.1	-	-	-	FD archives
Sri Lanka	Kendahena	1	0.1	8	10-17	776-438	12.4-18.7	17.7-27.3	-	-	12.0	-	-	-	FD archives
Sri Lanka	Kendahena	2	0.1	7	10-16	1025-348	12.6-17.8	19.5-26.3	-	-	8.7	-	-	-	FD archives
Sri Lanka	Kendahena	3	0.1	7	10-16	926-358	11.9-17.7	16.7-26.3	-	-	8.8	-	-	-	FD archives
Sri Lanka	Kendahena	4	0.1	8	10-17	667-219	13.5-20.3	19.9-29.5	-	-	7.1	-	-	-	FD archives
Sri Lanka	Kendahena	5	0.1	8	10-17	846-299	11.7-16.5	17.8-25.1	-	-	6.4	-	-	-	FD archives
Sri Lanka	Kendahena	6	0.1	8	10-17	796-328	12.5-21.6	17.3-29.8	-	-	12.0	-	-	-	FD archives
Sri Lanka	Kendahena	7	0.1	8	10-17	916-448	12.3-18.9	17.0-26.7	-	-	12.6	-	-	-	FD archives
Sri Lanka	Kendahena	8	0.1	8	10-17	567-289	14.4-20.7	21.6-29.8	-	-	9.7	-	-	-	FD archives
Sri Lanka	Kendahena	9	0.1	8	10-17	716-239	12.4-19.1	19.7-27.2	-	-	6.8	-	-	-	FD archives
Sri Lanka	Kendahena	10	0.1	8	10-17	896-299	13.8-19.8	19.6-28.2	-	-	9.2	-	-	-	FD archives
Sri Lanka	Kendahena	11	0.1	8	10-17	1214-517	13.2-18.0	20.9-25.1	-	-	13.2	-	-	-	FD archives
Sri Lanka	Kendahena	12	0.1	8	10-17	667-438	12.3-19.5	19.3-29.3	-	-	13.1	-	-	-	FD archives
Sri Lanka	Paspolakande	W2A	0.2	6	14-19	485-247	14.4-18.5	21.0-26.2	-	-	6.6	-	-	-	FD archives
Sri Lanka	Paspolakande	W2B	0.2	6	14-19	543-323	15.0-20.6	21.3-28.0	-	-	10.8	-	-	-	FD archives
Sri Lanka	Sundapola	-	-	2	27-30	-	36.0-40.0	-	-	-	-	-	-	-	Lushington (1921)

[a] Not all dbh and height measurements were taken at every remeasurement
[b] Additional trees were included in the sample plot at some remeasurements
[c] Maximum height or diameter measurements for the sample plot were recorded at the penultimate remeasurement, due to subsequent crown thinning.
[d] Values given are 'predominant' (pertaining to the 20 largest trees per hectare) rather than dominant
[e] Also Leitch (1991)
[f] Also SRD (1989)
[g] Also Wright (1968)
[h] (1992)

Appendix V Yield tables for mahogany stands

For abbreviations see Table 9.1.

Table V-A. Empirical / provisional yield table for three site classes from mahogany plantations in St. Lucia (Whitman, no date)

Age	H_d	N	G	D_m	$V_{7.6u}$*
\multicolumn{6}{c}{Site class 1}					
5	12.5	1050	12.0	10.2	89.5
10	18.9	716	25.2	19.2	181.6
15	23.2	506	33.5	27.4	237.6
20	26.5	408	36.7	33.5	276.9
25	29.2	346	37.4	36.8	292.3
30	40.6	309	36.7	40.1	300.7
35	34.1	277	35.3	42.7	306.3
40	36.0	247	34.0	44.2	307.7
45	37.8	235	31.9	45.7	307.0
		Site class 2			
5	10.4	1223	6.2	8.4	39.2
10	14.0	889	12.6	16.0	79.7
15	17.0	598	19.0	21.8	121.7
20	20.4	464	25.5	27.9	163.8
25	22.9	333	31.7	31.7	201.4
30	25.3	326	33.5	35.0	241.3
35	27.1	289	35.4	37.6	264.3
40	28.6	264	36.0	39.4	283.1
45	29.9	247	37.2	41.4	302.1
		Site class 3			
5	8.8	1383	6.8	7.6	33.6
10	11.3	1050	12.2	15.0	62.9
15	13.4	790	18.8	20.3	95.1
20	15.5	622	24.8	25.4	127.3
25	17.0	514	29.4	28.7	159.4
30	18.6	420	33.0	31.7	190.2
35	19.8	356	35.8	33.8	214.0
40	21.3	332	37.0	35.0	229.4
45	22.3	314	37.0	36.3	240.6

* Standing timber volume only   Reliability class: 2

Background information: The data on which the yield table is based cover trees 7 to 22 years old and 10 to 28 m tall. All predictions in the table from ages 22 to 45 are therefore based on extrapolation and should be treated with caution. No associated growth functions have been described.

Table V-B. Provisional yield table for three site classes from mahogany plantations in Indonesia (Anon. 1975)

Age	N	Site class 1			Site class 2			Site class 3		
		D_m	V_{MAI}	V	D_m	V_{MAI}	V	D_m	V_{MAI}	V
10	1070	17.6	16.7	176	12.4	9.8	98	10.1	5.0	50
15	1070	24.7	18.3	275	19.6	12.3	185	12.8	7.0	105
20	500	30.7	18.5	370	27.2	13.6	272	17.1	8.6	172
25	500	36.2	18.3	458	32.6	14.3	358	22.0	9.9	248
30	325	41.5	17.8	534	36.5	14.6	438	26.5	11.1	333
35	325	46.2	17.3	606	40.0	14.8	518	31.1	11.9	417
40	240	51.6	16.8	672	43.5	14.7	588	35.1	12.3	492
50	181	62.1	15.9	795	49.8	14.4	720	41.8	12.7	635
60	140	76.2	14.9	894	55.5	13.8	828	45.6	12.5	750

Reliability class: 1-2

Background information: Data were taken from 36 PSPs located at Telawa, Semarang, Gunuru, Kichul and Tasikmalaya, at an altitude of 150-600 m. No more information on growth data is given and growth functions are not described.

Table V-C. Empirical yield table for two site classes from mahogany plantations in Belize, assuming a constant stocking density of 300 stems ha^{-1} (Ennion 1996)

Age	H_d	D_m	G	V_{10}	V_{MAI}
Best site					
5	6.8	4.2	0.4	-	-
10	11.1	9.2	2.0	-	-
15	14.8	14.0	4.6	19.5	1.3
20	18.4	19.0	8.5	51.8	2.6
25	22.0	23.0	12.5	93.7	3.8
30	25.1	26.1	16.1	138.7	4.6
35	28.3	28.5	19.1	186.1	5.3
40	31.1	30.3	21.6	231.6	5.8
Worst site					
5	4.9	4.2	0.4	-	-
10	8.6	9.2	2.0	-	-
15	12.0	14.0	4.6	16.9	1.1
20	15.7	19.0	8.5	46.0	2.3
25	18.6	23.0	12.5	82.1	3.3
30	21.6	26.1	16.1	122.8	4.1
35	24.2	28.5	19.1	163.4	4.7
40	26.8	30.3	21.6	203.1	5.1

Reliability class: 2

Background information: See Appendix IV for description of raw data, Table 9.3 for height growth functions and Table 9.8b for the single tree volume tables from which stand volume estimates were made (the Sri Lanka equation was used). Yield tables for lower stocking (100 and 200 stems ha^{-1}, constant throughout the rotation) are available but have not been included because the relationship between tree age and tree volume appears to be independent of stocking (due to the lack of competition between trees). Volumes for lower stocking densities may be calculated by multiplying stated volume by a stocking fraction (new stocking/300).

Table V-D. Normal yield table for three site indices from mahogany plantations in the Philippines (Revilla et al. 1976a)

Age	N	Site index 20			Site index 25			Site index 30		
		D_m	H_d	V_{20u}*	D_m	H_d	V_{20u}*	D_m	H_d	V_{20u}*
10	1720	11.6	5.9	-	12.8	8.6	-	14.0	10.4	-
15	-	-	9.4	20.9	-	11.8	21.3	-	14.1	21.7
20	690	19.6	11.7	49.3	21.2	14.7	58.4	22.6	17.6	69.0
25	-	-	14.0	82.5	-	17.4	106.9	-	20.9	138.4
30	440	26.8	16.0	116.3	28.8	20.1	160.0	30.8	24.1	219.9
35	-	-	18.0	148.6	-	22.6	213.4	-	26.2	306.3
40	330	33.0	20.0	178.6	35.0	25.0	264.8	37.0	30.0	392.5
45	-	-	21.9	206.1	-	27.4	313.3	-	32.8	476.2
50	270	38.2	23.7	231.0	40.0	29.7	358.3	41.8	35.6	555.8
55	-	-	25.5	253.7	-	31.9	400.0	-	38.3	630.7
60	230	42.0	-	-	43.6	-	-	44.2	-	-

* Standing timber volume only   Reliability class: 2

Background information: See Table 9.3 for height growth functions. Volume was calculated using the equation: $V_{20} = 10^{(1.7348 - 6.6721/A + 0.053801S - 0.78406)}$. See Table 9.8b for single tree volume tables from which stand volume estimates were made. The data used to produce the yield table cover trees from 5 to 55 years old with a site index from 10 to 35 m (index age = 40 years). Yield predictions for additional site indices (10, 15 and 35) are described in Revilla (1976a), but since they were based on a small proportion (10% between them) of the sample they have been omitted here. The yield table has been verified for Makiling forest in Laguna, Piddig in Ilocos Norte, Pugo in La Union, Lagangilang in Agra, Santa Fe in Nueva Vizcaya, Cebu City and Minglanilia in Cebu, Dagohoy in Bohol, Bantay in Ilocos Sur. However, the yield table was produced and subsequent verification carried out using temporary sample plot data only.

Table V-E(i). Provisional yield table based on mean annual increment in volume from mahogany plantations in Fiji, assuming a 35 year rotation (AIDAB 1986)

Age	Operation	Timber yield (m³ ha⁻¹)			
		V_{MAI} = 5	V_{MAI} = 6	V_{MAI} = 7	V_{MAI} = 8
15	1st thin	15	20	20	25
25	2nd thin	25	35	45	50
35	final felling	135	155	180	205

Reliability class: 2   No background information available

Table V-E(ii). Provisional yield table based on mean annual increment in volume from mahogany plantations in Fiji, assuming a 50 year rotation (AIDAB 1986)

Age	Operation	Timber yield (m³ ha⁻¹)			
		V_{MAI} = 3.8	V_{MAI} = 4.5	V_{MAI} = 5.3	V_{MAI} = 6.0
15	1st thin	15	20	20	25
25	2nd thin	25	35	45	50
35	3rd thin	25	35	45	50
50	final felling	125	135	155	175

Reliability class: 2   No background information available

Table V-F. Normal yield table for four site classes from mahogany plantations in Martinique (Tillier 1995)

Age	H_d	N	H_m	Standing timber						Thinning and felling yields					V_7	Total yield		
				D_m	D_d	G	V_7	V_{20}	V_m	n	V_m	V_7	V_{20}	d_g		V_{20}	V_{MAI}	V_{CAI}
Site class 1																		
21	27.6	600	22.0	26.7	38.2	33.9	309	203	0.52	400	0.36	145	78	21.3	454	281	21.6	35.0
28	32.0	350	26.4	35.3	46.8	34.2	360	292	1.03	250	0.72	180	136	25.5	686	428	24.5	25.8
35	34.6	225	31.3	46.5	53.8	38.3	433	383	1.93	125	0.95	118	95	26.4	877	614	25.1	16.4
42	36.1	200	34.4	53.8	59.5	45.5	527	478	2.64	25	1.24	31	26	26.7	1002	735	23.9	10.6
50	37.0	-	35.8	58.3	65.9	53.4	613	561	3.07	200	3.06	613	561	58.3	1088	818	21.8	8.4
Site class 2																		
21	23.1	600	18.7	26.1	34.4	31.8	258	154	0.43	400	0.30	122	55	20.7	380	199	18.1	28.4
28	27.4	350	23.1	34.1	42.7	31.8	298	231	0.85	250	0.60	149	105	25.1	568	337	20.3	22.4
35	30.1	225	27.9	44.6	49.3	35.2	359	311	1.60	125	0.78	98	76	26.1	728	492	20.8	15.0
42	31.9	200	30.9	51.6	55.1	41.9	442	395	2.21	25	1.04	26	21	26.4	836	597	19.9	10.1
50	33.0	-	32.3	56.0	61.1	49.4	522	473	2.61	200	2.61	522	473	56.0	917	675	18.3	8.1
Site class 3																		
21	19.0	600	15.7	25.1	31.2	29.8	217	114	0.36	400	0.26	102	36	20.1	319	150	15.2	22.1
28	22.9	350	19.8	32.8	38.8	29.5	243	179	0.70	250	0.49	122	79	24.8	467	257	16.7	18.8
35	25.7	225	24.4	42.7	45.2	32.2	293	247	1.30	125	0.64	80	58	26.4	597	384	17.1	13.4
42	27.6	200	27.3	49.3	50.6	38.3	364	319	1.82	25	0.86	21	17	27.1	689	473	16.4	9.5
50	29.0	-	28.9	53.8	56.7	45.7	438	392	2.19	200	2.19	484	392	53.8	764	545	15.3	7.8
Site class 4																		
21	15.3	600	12.9	24.2	28.3	27.8	185	83	0.31	400	0.22	87.1	22	19.1	272	105	13.0	16.3
28	18.7	350	16.7	31.5	35.3	27.4	198	134	0.57	250	0.40	98.9	57	24.2	384	191	13.7	15.1
35	21.4	225	21.0	40.7	41.4	29.3	236	191	1.05	125	0.51	64.3	43	26.4	486	290	13.9	11.4
42	23.4	200	23.7	47.1	46.8	34.8	294	252	1.47	25	0.69	17.3	13	28.0	562	364	13.4	8.6
50	25.0	-	25.4	51.9	52.5	42.0	362	317	1.81	200	1.81	362	317	51.9	629	429	12.6	7.4

Reliability class: 1

Background information: See Tables 9.3 and 9.4 for height growth functions, Table 9.6 for diameter growth functions and Table 9.8b for single tree volume tables from which stand volume estimates were made.

Appendix VI Volume tables for mahogany trees

Not all of the following volume tables are directly comparable; refer to notes under each table for precise definitions of height and volume.

Table VI-A. Single entry volume tables for mahogany from various countries (after Orden 1956, Wescom 1979, Funes *et al.* 1983 and Mayhew in press, b)

dbh (cm)	Timber volume (m³)				
	Honduras	Sri Lanka	Fiji [a]	Fiji [b]	Philippines
10	-	0.02	0.03	0.01	-
15	-	0.08	0.09	0.07	-
20	-	0.19	0.18	0.14	0.13
25	-	0.35	0.29	0.24	0.20
30	0.18	0.56	0.42	0.35	0.30
35	0.28	0.83	0.58	0.49	0.42
40	0.39	1.15	0.76	0.65	0.56
45	0.54	1.51	0.96	0.83	0.73
50	0.72	1.94	-	-	0.93
55	0.92	2.41	-	-	1.15
60	1.16	2.93	-	-	1.39
65	1.44	3.51	-	-	1.66
70	1.75	4.14	-	-	1.96
75	2.11	-	-	-	2.28
80	2.50	-	-	-	2.62

[a] Trees on good-average sites
[b] Trees on poor sites
Honduras: volume of main stem to first branch or top diameter of 30 cm underbark
Sri Lanka: volume of main stem to 10 cm top diameter overbark
Fiji: volume of main stem to first branch or 10 cm top diameter overbark
Philippines: volume of main stem to first branch or 15 cm (?) top diameter underbark

Table VI-B. Double entry volume table for mahogany from Honduras (Funes *et al.* 1983)

dbh (cm)	Timber volume (m³) by height				
	5 m	10 m	15 m	20 m	25 m
30	0.28	0.45	0.59	0.72	0.84
35	0.38	0.61	0.80	0.98	1.14
40	0.49	0.79	1.04	1.27	1.48
45	0.62	0.99	1.31	1.60	1.87
50	0.76	1.22	1.62	1.97	2.30
55	0.91	1.47	1.95	2.38	2.77
60	1.09	1.75	2.31	2.82	3.29
65	1.27	2.05	2.71	3.30	3.85
70	1.47	2.37	3.13	3.82	4.45
75	1.68	2.71	3.59	4.38	5.10
80	1.91	3.08	4.08	4.97	5.79

Height = height to first branch or 30 cm top diameter
Volume = volume of main stem to first branch or 30 cm top diameter underbark

Table VI-C. Double entry volume table for mahogany from St. Lucia (Andrew 1994)

dbh (cm)	Timber volume (m³) by height					
	10 m	15 m	20 m	25 m	30 m	35 m
10	0.05	0.07	0.08	0.09	0.11	0.12
15	0.09	0.12	0.15	0.18	0.21	0.24
20	0.13	0.19	0.24	0.30	0.35	0.41
25	0.20	0.28	0.37	0.45	0.54	0.62
30	0.27	0.39	0.51	0.64	0.76	0.88
35	0.36	0.52	0.69	0.86	1.02	1.19
40	0.46	0.68	0.89	1.11	1.33	1.55
45	0.58	0.85	1.13	1.40	1.68	1.95
50	0.70	1.04	1.38	1.72	2.06	2.40
55	0.85	1.26	1.67	2.08	2.49	2.90
60	1.00	1.49	1.98	2.47	2.96	3.45
65	1.17	1.75	2.32	2.89	3.47	4.04

Height = total height   Volume = volume of main stem to 9 cm top diameter underbark

Table VI-D. Double entry volume table for mahogany from Sri Lanka (Mayhew in press, b)

dbh (cm)	Timber volume (m³) by height						
	10 m	15 m	20 m	25 m	30 m	35 m	40 m
10	0.01	0.02	0.02	0.02	0.03	0.03	0.03
15	0.06	0.08	0.10	0.12	0.14	0.16	0.17
20	0.13	0.17	0.21	0.25	0.29	0.33	0.37
25	0.21	0.28	0.35	0.42	0.49	0.56	0.63
30	-	0.42	0.52	0.63	0.73	0.83	0.94
35	-	0.58	0.73	0.87	1.01	1.16	1.30
40	-	0.77	0.96	1.15	1.34	1.53	1.72
45	-	-	1.23	1.47	1.72	1.96	2.20
50	-	-	1.53	1.83	2.13	2.44	2.74
55	-	-	1.85	2.22	2.59	2.96	3.33
60	-	-	-	2.66	3.10	3.54	3.98
65	-	-	-	3.13	3.64	4.16	4.68
70	-	-	-	3.63	4.24	4.84	5.44

Height = total height   Volume = volume of main stem to 10 cm top diameter overbark

Table VI-E. Double entry volume table for mahogany from the Philippines (Revilla et al. 1976b)

dbh (cm)	Timber volume (m³) by height				
	5 m	10 m	15 m	20 m	25 m
20	0.12	-	-	-	-
25	0.18	-	-	-	-
30	0.26	0.43	0.57	-	-
35	0.36	0.58	0.78	-	-
40	0.47	0.76	1.01	1.24	1.45
45	0.59	0.97	1.29	1.58	1.84
50	0.74	1.20	1.59	1.95	2.28
55	0.89	1.45	1.93	2.36	2.76
60	1.06	1.73	2.30	2.82	3.30
65	1.25	2.03	2.71	3.31	3.87
70	1.45	2.36	3.14	3.85	4.50
75	1.67	2.72	3.61	4.42	5.17
80	1.90	3.09	4.11	5.04	5.89

Height = height to first branch or 20 cm top diameter
Volume = volume of main stem to first branch or 20 cm top diameter underbark

Table VI-F. Double entry volume table for mahogany from the Philippines (Revilla *et al.* 1976b)

dbh (cm)	Timber volume (m³) by height					
	5 m	10 m	15 m	20 m	25 m	30 m
20	0.11	-	-	-	-	-
25	0.16	-	-	-	-	-
30	0.23	0.40	0.55	-	-	-
35	0.31	0.54	0.74	-	-	-
40	0.41	0.70	0.95	1.19	1.42	-
45	0.51	0.87	1.20	1.50	1.78	-
50	0.63	1.07	1.47	1.83	2.18	2.51
55	0.75	1.29	1.76	2.21	2.62	3.02
60	0.89	1.52	2.09	2.61	3.10	3.58
65	1.04	1.78	2.44	3.05	3.62	4.18
70	1.20	2.06	2.81	3.52	4.18	4.82
75	1.37	2.35	3.22	4.02	4.78	5.51
80	1.55	2.66	3.65	4.56	5.42	6.24

Height = height to 20 cm top diameter
Volume = volume of main stem to 20 cm top diameter underbark

Table VI-G. Double entry volume table from Guadeloupe (Soubieux 1983)

dbh (cm)	Timber volume (m³) by height				
	5 m	10 m	15 m	20 m	25 m
10	0.02	0.03	0.04	0.05	0.05
15	0.06	0.08	0.11	0.13	0.14
20	0.11	0.15	0.20	0.25	0.28
25	0.17	0.23	0.32	0.40	0.45
30	0.24	0.34	0.46	0.58	0.67
35	0.32	0.46	0.64	0.79	0.92
40	0.41	0.61	0.83	1.04	1.22

Height = total height
Volume = volume to 7 cm top diameter overbark

Index

Acacia
   auriculiformis, 67
   mangium, 67, 132
Acrocerops auricilla, 43
Africa, 70, 125, 140
Agathis sp., 72
agroforestry, 56, 58-65, 69, 79, 134, 171
Albizia
   falcataria, 65, 67, 68, 77
   flacata, 64
   sp., 64, 67, 68, 72, 74, 77
ambrosia beetle, 42, 74, 75, 94, 124, 141-143, 144, 151
Anderson groups, 59
Anthocephalus
   chinensis, 68
   renalis, 138
Antigua, 19
Antilles, 3, 88, 159, 162, 164
ants, 146
Apanteles leptanus, 138
Armillaria mellea, 145, 147
Artocarpus integrifolia, 21, 57, 64, 66, 93, 133, 160, 163, 164
Aspidospermum megalocarpum, 77
Attalea cohune, 12
Australia, 19, 78, 125, 145
avenues, 17, 18, 19, 27
Azadirachta
   excelsa, 128
   indica, 128

Bacillus thuringensis, 140
Bagassa guianensis, 63
balsa, 132, 133
bananas, 62, 63, 132

Bangladesh, 19
bare-rooted plants (see planting stock)
bark, 13, 38, 64, 67, 72, 73, 83, 93, 119, 124, 144, 145, 146, 147, 148, 150, 151
basal area, 9, 11, 20, 85, 87, 89, 90, 91, 96, 98, 99, 108
beating up, 56, 60
Beauvaria
   bassiana, 140
   tenella, 140
Belize, 2, 7, 9, 10, 11, 13, 14, 15, 18, 19, 22, 23, 26, 28, 49, 52, 55, 56, 59, 61, 62, 77, 78, 88, 95, 100, 103, 106, 107, 111, 113, 114, 115, 116, 120, 124, 131, 132, 133, 134, 135, 138, 143, 146, 149, 153, 155, 157, 158, 160, 167, 169
Bertholettia excelsa, 63
blue mahoe, 53, 93, 164
Bolivia, 4, 7, 8, 9, 12, 13, 14, 15, 122, 124, 153
Bolotia sp., 78
Borneo, 20
Botryodiplodia theobromae, 41, 148
Brachiaria humidicola, 67
Brazil, 2, 3, 4, 7, 15, 19, 26, 30, 35, 44, 60, 63, 69, 76, 103, 132, 133, 134, 171
British Honduras (see also Belize), 18, 59, 155
Buchanavia tetrophylla, 53

Calophyllum brasiliense, 77
capsules, 8, 9, 29, 30
Caribbean, 3, 18, 23, 26, 53, 59, 125, 138, 139, 140, 149

220

Cassia simea, 64, 133
Cecropia peltata, 134
Cedrela odorata, 14, 60, 61, 63, 72, 129, 132
Cedrelinga catenaeformis, 72
Central America, 3, 6, 16, 23, 27, 29, 41, 122, 150, 169
Centro Agronómico de Investigación y Enseñanza (CATIE), 1, 24, 137
Cercospora subsessilis, 148
Chile, 125
Chimanes, Bolivia, 9, 12, 13, 14, 153
Chiquibul, Belize 10, 60, 78, 157
Chloroxylon swietenia, 164
cleaning, 60, 70, 74, 77, 82, 93, 96, 103, 104, 135, 141, 142, 143, 154, 155, 156, 167
cocoa, 21, 62, 79
coffee, 19, 53, 59, 61, 62, 79
Coleoptera, 141
Colombia, 25, 139
Columbia River, Belize, 23, 55, 59, 60, 100, 106, 134, 155
container plants (see planting stock)
conversion
    line planting, 69-76, 78, 85, 94, 96, 109, 128, 131, 132, 133, 143, 171
    underplanting, 69, 76-78, 79
coppicing, 37, 38, 45, 82, 158
Coptotermes sp., 143
Cordia
    alliodora, 63, 65, 133
    goeldiana, 63
Corticium koleroga, 42
Costa Rica, 1, 2, 3, 10, 19, 22, 23, 24, 128, 132, 133, 137, 146
creeper cutting, 81, 143, 156, 157
crown diameter, 74, 88, 89, 90, 96
Cuba, 19, 125, 148
Cunicula paca, 15
Cusenta chinensis, 82
cuttings, 37, 38, 45
cyclones (see hurricanes)

Dacrydium sp., 72
Dalbergia latifolia, 64
Dasyprocta punctata, 15
Dendropthoe falcata, 146

Desmodium intortum, 63
diameter
    distribution, 12, 13, 15, 108, 161
    dominant, 98, 112, 120
    mean, 60, 89, 90, 97, 98, 99, 100, 104, 108-111, 112, 120
Diaprepes abbreviatus, 42, 146
direct seeding, 54-56, 59, 151, 169
diseases, 139, 140, 145-148
disturbance, 10, 12, 13, 14, 16, 48, 142, 143, 154, 157, 160, 167, 169
Dominican Republic, 19
Dysercus longiclaris, 146

ecological range, 7-8, 14, 15, 47
Ecuador, 7
Egchiretes nominus, 146
Endospermum sp., 72
enrichment planting (see also conversion line planting), 27, 69, 78, 79, 132, 137, 142
Erythrina
    poeppigigiana, 128
    sp., 128
Eucalyptus
    deglupta, 67
    sp., 64, 66, 67

fecundity, 9
felling
    clear, 168
    cycle, 153, 154, 157
    group, 165, 166, 169
    restrictions, 153, 154, 167
    selective, 21, 157, 160, 167
fertilising, 38, 82-83, 134, 135, 140
Ficus
    elastica, 148
    sp., 73, 148
Fiji, 2, 18, 19, 22, 23, 24, 25, 26, 27, 28, 29, 30, 34, 35, 36, 38, 39, 40, 41, 42, 43, 44, 45, 48, 49, 50, 51, 56, 65, 71, 72, 73, 74, 75, 81, 82, 83, 88, 89, 91, 94, 95, 96, 102, 103, 105, 106, 107, 109, 113, 117, 119, 120, 123, 124, 125, 141, 142, 143, 144, 145, 147, 149, 152, 159, 160, 162, 169
fire, 13, 14, 15, 70, 74, 150-151, 169

Flavopimpla latiannulata, 138
flooding, 12, 15, 169
Florida, 3, 125
flowers, 9, 83
Flueggea flexuosa, 75
Freshwater Creek, Belize 55, 56, 149, 155
fruits (see also capsules), 3, 8, 14, 28, 33, 83
fungicide, 29, 41
Fusarium sp., 41

Galoa, Fiji, 71, 72, 74, 83
genetic
   improvement, 24-25, 26
   variation, 24, 25, 128, 129, 132
germination (see seed germination)
girdling, 72, 73, 77, 79, 143, 147, 148, 156
Glomerella cingulata, 148
Gmelina arborea, 60, 65, 67, 133
grafting, 25
Grenada, 19, 81, 128, 138
growth
   functions, 97, 101, 102, 103, 104, 105, 106, 107, 108, 109, 110, 111, 112, 113, 114, 121
   rate, 10, 12, 23, 34, 40, 48, 50, 51, 52, 53, 57, 59, 60, 63, 70, 73, 77, 78, 80, 81, 82, 83, 85, 86, 88, 91, 95, 96, 99, 100, 103, 104, 105, 109, 111, 120, 123, 124, 135, 137, 149, 154, 155, 164, 167, 170
Guadeloupe, 19, 22, 26, 47, 49, 51, 58, 76, 77, 81, 101, 103, 108, 109, 112, 114, 118, 119, 120, 142, 152
Guajataca, Puerto Rico, 21, 51
Guam, 19
Guatemala, 14, 18, 22, 23, 56, 125, 133, 149
Guyana, 19

Haiti, 19
heartwood, 122, 123, 124, 145
height
   dominant, 68, 97, 98, 100, 101, 105-108, 109, 112, 120

mean, 54, 98, 100, 104-105, 108, 109, 119, 120
Heliocarpus donnel, 60
Heliopittis antonii, 42
herbicides, 72, 73, 79
herbivory, 15, 43, 146
Heteropsylla cubana, 65, 68
Heterorhabditis sp., 145
Hexacolus guyanaensis, 142
Hibiscus elatus, 21, 53, 64, 93, 133, 164
Holland, 22
Honduras, 2, 14, 18, 19, 22, 23, 24, 25, 26, 43, 44, 49, 50, 59, 75, 76, 77, 78, 83, 84, 85, 101, 103, 112, 117, 118, 119, 120, 123, 132, 133, 146, 155
hurricanes, 13, 14, 15, 19, 23, 24, 27, 55, 95, 142, 143, 149, 150, 151, 153, 164, 166, 167
hybrids of mahogany, 3, 10, 50, 52, 85, 94
Hypothenmus eruditus, 142
Hypsipyla
   grandella, 14, 24, 125, 126, 128, 138, 139, 140
   robusta, 125, 138, 139, 140
   sp. (see also shoot borer), 14, 24, 40, 43, 46, 125, 126, 127, 128, 129, 130, 131, 133, 134, 135, 137, 138, 139, 140, 170

Imperata
   cylindrica, 82
   sp., 58, 81, 82
India, 17, 18, 19, 22, 27, 29, 42, 64, 75, 82, 125, 138, 146, 147, 148
Indonesia, 17, 18, 20, 22, 26, 27, 29, 39, 42, 48, 49, 50, 52, 58, 61, 64, 65, 86, 87, 88, 90, 91, 92, 94, 95, 111, 113, 114, 115, 120, 125, 142, 146, 147, 148, 151, 152, 159, 162, 169
Inga
   edulis, 67, 69, 133
   sp., 63, 67, 69, 133
insecticides, 10, 32, 42, 137
Instituto Nacional de Investigaciones Forestales y Agropecuarias (INIFAP), 24

jak, 64, 66, 93, 164
Jamaica, 20, 22
Java, 20, 22, 39, 42, 49, 51, 57, 58, 61, 95, 132, 133

Kegalle, Sri Lanka, 49, 117, 118
Khaya sp., 3
Kolombangara, Solomon Islands, 17, 21, 50, 52, 61, 72, 74, 84, 124, 134
Kolombangara Forest Products Ltd. (KFPL), 17, 84
Kurunegala, Sri Lanka, 117, 118

Lancetilla, Honduras, 19, 23, 24, 25, 117, 118
land races, 24
Lantana
  camara, 82
  sp., 80, 82
Las Caobas, Mexico, 6, 11, 154
leaf miners, 82
leaves, 3, 4, 9, 13, 15, 25, 29, 35, 41, 42, 44, 46, 47, 82, 125, 146, 148, 149, 151, 156
Lepidoptera, 14, 125
Leucaena
  glauca, 64
  leucocephala, 65, 68
  sp., 64, 65, 66, 68
lifting, 39, 40, 43, 46
line planting (see conversion line planting)
litter layer, 12, 160
logging (see felling)
Luquillo, Puerto Rico, 21, 48, 49, 50, 56
Luzon, 18, 39, 49

Macadamia sp., 129
maize, 59, 60, 61, 62, 63, 79
Malaysia, 2, 18, 20, 22, 30, 39, 56, 57, 58, 64, 65, 75, 76, 81, 86, 94, 132, 134, 142, 146
Manilkara bidentata, 53
Martinique, 20, 22, 26, 30, 43, 48, 49, 50, 53, 57, 58, 61, 84, 86, 87, 88, 90, 91, 92, 95, 106, 107, 108, 111, 112, 113, 114, 115, 120, 124, 132, 134, 136, 152, 158, 170
Mauritius, 20
mayflower, 60
mean annual increment (see also yield prediction), 95, 98, 99, 103, 114, 116, 120
Meliaceae, 2, 3, 14, 31, 32, 37, 38, 40, 44, 122, 124, 128, 129, 131, 133, 136
Merremia
  sp., 68, 75, 80, 81, 82
  umbellata, 82
Metarhizum anisopliae, 140, 145
Mexico, 2, 3, 7, 9, 10, 11, 13, 14, 15, 20, 22, 23, 24, 25, 27, 28, 36, 49, 57, 58, 63, 76, 78, 80, 81, 95, 103, 105, 109, 133, 134, 149, 154, 160
Micosphaerella sp., 148
micropropagation, 38
Mikania
  micrantha, 82
  scandens, 80
  sp., 80, 82
moho, 60
Montserrat, 20
Myanmar, 20
Myristica castaneifolia, 143

natural distribution (see also ecological range), 5, 6
natural forest improvements, 11, 131, 155, 156
natural forest management, 1, 2, 153-158 167, 169
natural regeneration, 10, 11, 12-14, 16, 21, 36, 83, 86, 93, 143, 144, 150, 151, 152, 153, 154, 156, 157, 158-166, 167, 169, 171
Neotermes
  papua, 144
  samoanus, 144
Nicaragua, 18, 20, 22, 23, 103
Nigeria, 20
Nukurua, Fiji, 34, 106, 117, 144
nursery techniques, 34-46

Ochroma limonensis, 60

Pacific, 3, 17, 18, 23, 26, 27, 125, 140, 149
Panama, 20, 22, 23, 103
Panicum maximum, 63
Pellicularia sp., 41
Peltophorum pterocarpa, 64
Peninsular Malaysia, 20, 22, 58, 64
Peru, 7, 9, 20, 22, 27, 29, 39, 72, 103, 125, 131, 132, 137, 148
Perum Perhutani, 20
pests (see also *Hypsipyla*), 15, 24, 38, 141-148
Petén, 56
Phanerotoma sp., 138
Phellinus
  fatuosus, 147
  noxius, 24, 65, 68, 70, 145
phenology, 9
Philippines, 2, 17, 18, 20, 22, 23, 25, 26, 27, 28, 29, 30, 34, 35, 39, 41, 42, 48, 49, 50, 52, 57, 58, 62, 65, 67, 81, 83, 85, 88, 95, 102, 106, 107, 111, 113, 114, 117, 118, 119, 120, 123, 134, 148, 149, 150, 152, 159, 169
Photorhabdus sp., 145
pine, 65, 149, 151
pin-holes, 124
Pinus
  caribaea, 24, 65, 133
  merkusii, 64
Plan Piloto Forestal (PPF), 154
plantation
  areas, 17, 19, 20, 21
  establishment, 2, 23, 54-79, 140, 164, 171
  maintenance, 60, 80-96, 127
  mixed, 57, 58-69, 79, 85, 93, 104, 129, 131, 133, 136, 164, 170
  pure, 57-58, 84, 85-92, 93, 133, 164
planting density (see stocking density)
planting stock
  bare-rooted, 39, 43, 44-45, 46
  container, 31, 32, 34, 35, 39, 40, 43
  oversized, 44, 46
Platypodidae, 141, 142
Platypus gerstaeckeri, 142
plus trees, 25, 28, 161
poisoning, 70, 72, 74, 77, 82, 143, 147

polak, 60
pollination, 8, 9
Pratylenchus coffeae, 147
pricking out, 35
Procryptotermes sp., 144
progeny tests, 23-24, 25
Programme for Belize, 157
protection (see also shoot borer control), 141-151
provenance trials, 21, 23-24, 25, 26, 47, 51, 130, 170
pruning, 59, 60, 61, 62, 70, 74, 83-84, 131, 134, 136, 140, 142, 143, 154, 170
Puerto Rico, 2, 9, 18, 21, 22, 23, 24, 25, 27, 28, 29, 40, 42, 48, 49, 50, 51, 52, 56, 58, 61, 70, 72, 73, 75, 76, 77, 80, 81, 82, 85, 91, 94, 103, 123, 125, 132, 133, 142, 146, 148
Pyralidae, 14, 125
Pythium splendens, 148

Quassia amara, 128
Queensland, 19, 78
Quintana Roo, Mexico, 6, 7, 9, 10, 11, 13, 24, 36, 49, 78, 133, 134, 154

Ramphastos sp., 15
reliability class, 104, 105, 106, 108, 109, 112
Rhizoctonia
  solani, 41
  sp., 41, 148
Rio Abajo, Puerto Rico, 21, 48
Rio Bravo Conservation and Management Area, 7, 157, 158
rodents, 146
root
  collar, 25, 40, 41, 42, 44, 145, 147
  competition, 11, 160
  decay, 24, 147, 148
  pruning, 39-40, 149
rotation length, 66, 79, 88, 91, 94-95, 96, 108, 114, 124, 162, 170

Sabah, 20, 22, 58, 75, 133, 137
Sabal excelsa, 12
St. Croix, 23

St. Lucia, 2, 21, 28, 34, 39, 52, 58, 61, 64, 91, 93, 106, 107, 111, 113, 114, 118, 119, 120, 132, 133, 134, 138, 148, 159, 160, 164, 165, 166, 167
sapwood, 122, 123, 124, 145
Sarawak, 22, 58, 65, 81
Schizolobium amazonicum, 69
Scleria sp., 80
Sclerotium sp., 41
Scolytidae, 141, 142
Securinega
  flexuosa, 65, 68, 133
  samoana, 67
seed
  bank, 15
  collection, 2, 27-28, 30
  dispersal, 9, 154
  germination, 10, 11, 15, 28, 30, 33, 35, 36, 37, 38, 39, 45, 54, 55, 56, 59, 79, 129, 154, 162
  origins, 17-23, 26, 99
  predation, 10, 15
  preparation, 29-30
  production, 25, 27, 33, 86, 156, 160
  storage, 32
  trees, 9, 12, 94, 151, 153, 154, 157, 162, 163, 164, 167, 169
  viability, 9, 27, 30-33, 55, 56, 160
Setaria sphaceleta, 63
Seychelles, 21
shelterwood, 76, 77, 79, 93, 131, 132, 133, 146, 155, 156, 158, 160, 162-164, 166, 168, 171
shoot borer
  attack, 2, 17, 23, 24, 38, 43, 54, 57, 58, 59, 60, 61, 63, 66, 67, 68, 70, 76, 78, 80, 82, 84, 93, 96, 105, 126, 127, 128, 130, 131, 132, 133, 134, 135, 136, 137, 140, 152, 156, 160, 164, 170
  control, 125-140
  species (see also *Hypsipyla*), 125
Sierra Leone, 21
Silk Grass, Belize, 11, 55, 60, 134, 155, 156
silvicultural systems
  clear-cutting, 158, 160, 167, 169, 171

  group selection, 150, 160, 164-166, 168, 171
  shelterwood, 158, 160, 162-164, 168
  single tree selection, 160-162, 166
Singapore, 18, 22
site
  class (see site quality)
  index, 98, 105-108, 113, 120
  quality, 95, 99, 100, 104, 105, 108, 109, 111, 112, 113, 114, 115, 116, 120, 130, 142
  selection, 47-53, 128, 136, 140, 150, 151, 170
soil
  erosion, 20, 47, 52, 53, 73, 74, 164, 166, 167
  pH, 48, 50
  types, 7, 12, 21, 30, 47, 48-52, 53, 132, 160, 169
Solomon Islands, 2, 17, 18, 21, 22, 27, 29, 35, 39, 40, 42, 43, 44, 49, 50, 52, 60, 61, 63, 65, 67, 68, 72, 73, 74, 75, 80, 81, 84, 94, 95, 101, 102, 124, 125, 132, 133, 134, 136, 143, 146, 159
Sorghum sp., 63
South America, 2, 6, 14, 17, 26, 125, 140
South Asia, 26
South-East Asia, 125, 149
Sri Lanka, 2, 18, 21, 22, 25, 26, 27, 28, 44, 48, 49, 52, 57, 64, 65, 66, 67, 77, 78, 80, 82, 85, 86, 87, 88, 89, 90, 92, 93, 94, 95, 107, 117, 118, 119, 120, 132, 133, 143, 145, 146, 147, 150, 151, 152, 159, 160, 162, 163, 164, 166, 167
stem decay, 148
Sterculia caribaea, 134
stocking density, 15, 54, 55, 59, 80, 85, 88, 89, 90, 91, 92, 94, 96, 97, 99, 100, 102, 103, 108, 109, 111, 113, 114, 120, 148, 154, 156
stooling, 37
Sumatra, 20, 132
Surinam, 21
Swietenia
  humilis, 3, 4, 10, 23, 32, 125

mahagoni, 3, 4, 17, 25, 52, 122, 124, 125
sp., 1, 2, 3, 4, 8, 9, 14, 15, 17, 21, 23, 37, 38, 50, 52, 122, 124, 129, 132, 133

Tabasco, Mexico, 95, 105, 109
Tabebuia pentaphylla, 60
Taiwan, 21
Taiwan Forestry Research Institute, 21
tap roots, 13, 35, 39, 45, 46, 56, 149, 151
taungya, 55, 56, 59, 60, 61, 135, 164
teak, 20, 53, 57, 64, 75, 82
Tectona grandis, 64, 66, 75
Terminalia calamansanai, 65, 68
termites, 124, 143-145, 151
Tetrastichus spirabilis, 138
Thailand, 21, 22
Theobroma grandiflorum, 63
thinning
  crown, 91, 93, 96
  cycle, 93, 164
  liberation, 60, 157
  low, 91
  regimes, 27, 79, 85-94, 96, 108, 109, 111, 114, 115, 116, 120, 164, 166
  selective, 91, 93, 96, 164, 170
  systematic, 60, 164
timber quality, 95, 122-124
Tobago, 21
Toledo, Belize, 49, 55, 59
Tonga, 21, 22, 23
Toona
  ciliata, 67
  sureni, 57
top diameter, 98, 114, 115, 116, 117, 118, 119, 120

Trema orientalis, 64, 133
Trichogramma sp., 139
Trichogrammatoidea robusta, 138

Uganda, 21
underplanting (see conversion underplanting)
United Fruit Company, 19, 23, 83

Vanuatu, 21, 22, 24, 25, 30, 52, 54, 63, 65, 101, 105, 109, 125, 146, 147, 149
Venezuela, 7, 58, 137, 148
Veracruz, Mexico, 76, 81
Virgin Islands, 22
Virola merondonis, 77
Vitex pinnata, 164
Viti Levu, Fiji, 19, 105, 109
Vochysia
  koschnyia, 77
  maxima, 63
volume estimation, 108, 114-120

weeding, 34, 38, 43, 55, 56, 61, 70, 74, 75, 77, 79, 80-82, 83, 95, 104, 140, 167
weevil, 42, 146
Western Samoa, 22, 26, 35, 39, 67, 73, 75, 81, 94, 123, 125, 149, 150

Xenorhabdus sp., 145
Xyleborus
  abruptoides, 43
  caffeae, 43
  morstattii, 142
Xylosandrus compactus, 142

yield prediction, 97, 113-114